THE DEADLY RISE OF **ANTI-SCIENCE**

OTHER BOOKS BY PETER J. HOTEZ

Forgotten People, Forgotten Diseases: The Neglected Tropical Diseases and Their Impact on Global Health and Development, 2008 (3rd ed., 2021)

Blue Marble Health: An Innovative Plan to Fight Diseases of the Poor amid Wealth, 2016

Vaccines Did Not Cause Rachel's Autism: My Journey as a Vaccine Scientist, Pediatrician, and Autism Dad, 2018

Preventing the Next Pandemic: Vaccine Diplomacy in a Time of Anti-science, 2021

Peter J. Hotez, MD, PhD

THE DEADLY RISE OF **ANTI-SCIENCE**

A Scientist's Warning

Johns Hopkins University Press
Baltimore

© 2023 Johns Hopkins University Press
All rights reserved. Published 2023
Printed in the United States of America on acid-free paper
9 8 7 6 5 4 3 2 1

Johns Hopkins University Press
2715 North Charles Street
Baltimore, Maryland 21218
www.press.jhu.edu

Library of Congress Cataloging-in-Publication Data is available.

ISBN 978-1-4214-4722-3 (hardcover)
ISBN 978-1-4214-4723-0 (ebook)

A catalog record for this book is available from the British Library.

Special discounts are available for bulk purchases of this book. For more information, please contact Special Sales at specialsales@jh.edu.

To the world's scientists under threat from authoritarian leaders who seek to intimidate us.

And to Drs. Mwelecele Ntuli Malecela, Donald McManus, Joe Cook, Paul Farmer, and Philip K. Russell, five remarkable global health heroes and humanitarians whom we lost too soon; to Joe Pava, my close friend since childhood, and Dr. Jim Silverblatt, my Yale College roommate; and to my inspirational uncle, Dr. Irving H. Goldberg, and my beautiful and loving mom, Jean (Goldberg) Hotez. The tragedy of losing them during the writing of this book leaves a huge, gaping hole.

A special thank-you to the Houston Police Department, Texas Medical Center Police Department, and the security forces at Texas Children's Hospital and Baylor College of Medicine for keeping my family safe. Thank you to the Anti-Defamation League Southwest for its advice and help on rising anti-Semitic attacks and aggression.

A known enemy is already half disarmed.

—LOUIS PASTEUR

My fate was, in a sense, exceptional. It is not false modesty but the desire to be precise that prompts me to say that my fate proved greater than my personality. I only tried to keep up with it.

—ANDREI SAKHAROV

Contents

Preface xi

1 | An Army of Patriots Turns against the Scientists 1

2 | Health Freedom Propaganda in America 24

3 | Red COVID 41

4 | An Anti-science Political Ecosystem 64

5 | A Tough Time to Be a Scientist 87

6 | The Authoritarian Playbook 105

7 | The Hardest Science Communication Ever 122

8 | Southern Poverty Law Center for Scientists 142

Literature Cited 163

Index 209

| Preface

This was not an easy book to write because it recounts both painful personal experiences and the sufferings of others. For the past two decades, I have been a target of anti-vaccine activists who started in grassroots groups before gaining strength and power through well-organized, well-funded, and highly aggressive organizations. They alleged initially that vaccines cause autism but later expanded their accusations to make far-reaching yet unsubstantiated claims about vaccines and how they cause a variety of illnesses.

As a result of the COVID-19 pandemic, we learned that this overreach was only the beginning. Now such activism goes beyond vaccines to attack multiple facets of the biomedical sciences in the United States. It has evolved into an all-encompassing program embraced by a US political system on the right acting out strong libertarian or extreme conservative beliefs. Increasingly, the far-right extremists have adopted what I consider anti-science activism, turning it into a new form of aggression against both science and scientists. Anti-science politics is especially strong in Texas and other southern US states, but its tentacles now reach throughout the nation.

I have seen a new and expanded level of anti-science aggression rising in America, which pervades state legislatures and the US federal government, including Congress and the judiciary, as well as the conservative media. In this book I share my fears that anti-science has become a dangerous social force that threatens both our national security and global stature as a nation renowned for its research institutions and universities. I explain how and why anti-science has started to globalize, extending into Canada and western Europe, and even to low- and middle-income countries in Africa and elsewhere.

Some are quick to blame the Trump White House for turning the United States into an "anti-vax nation" in the years leading up to the pandemic and after COVID-19 first emerged in the United States; however, I believe this is simplistic and inaccurate. While President Donald J. Trump may have helped to crystallize some of the factors that promote anti-science in America, this development extends past his persona, actions, and public statements. Rather, anti-science politics began before Trump took office and has worsened since he left. It cuts across state and federal governments as well as civil society, and it promotes extreme libertarian viewpoints—its proponents termed by some as "Middle American Radicals" or the "New Right"—to weaponize an agenda of brazen anti-science rhetoric and propaganda.

The anti-science political machine has become a monster of significant proportions, and it won't end with COVID-19. Instead, COVID-19 is a catalyst for a new and scary beginning. Something terrible has happened to broaden and intensify public rejection of vaccines and other biomedical innovations in the United States, and now globally. In this book, I tell a very personal story of going up against the monster. In 2021–22, more than 200,000 unvaccinated Americans needlessly lost their lives to COVID-19 because they refused vaccines. I will explain what I see as the progression that led these 200,000 Americans to become, first, targets of and, ultimately, victims of anti-science aggression.

I detail how the anti-science component of far-right authoritarianism is now accelerating in the United States and how it manifests as regular attacks against scientific concepts, scientific institutions, and prominent American scientists. These extreme views originate from at least a half-dozen members of the US House of Representatives and Senate as well as several state governors and federal judges. Their attacks are amplified nightly on Fox News and other extreme conservative news outlets and podcasts. They are also supported by a cadre of contrarian intellectuals or pseudointellectuals from far-right think tanks and other institutions. In this sense, the US government, media, and the intellectuals who give academic cover to these viewpoints are waging a complicated and multifaceted war. It is one without an obvious end.

In this book, I also point out similarities to past and current authoritarian regimes that attack both science and scientists. They include examples from history, such as the Soviet Union under Stalin, and current nations and recent leaders, including Hungary under Viktor Orbán and Brazil under its former president, Jair Bolsonaro.

I'm fearful for the future of our country and its scientific institutions. I'm worried for my scientific colleagues and for my personal safety and that of my own family. It is essential that we somehow defeat the anti-science monster or at least find a way to keep it in check, walled off from the American scientific enterprise. Toward that end, I outline a plan to begin to restore the American people's trust in science and scientists. This book is my focused account of the events affecting biomedical sciences and scientists, although it also compares how the community of climate scientists and those in other scientific fields have dealt with similar attacks on their work.

Just talking about the politics of anti-science aggression is itself difficult. Our training as physician-scientists says that we should not discuss the politics of Republicans versus Democrats or conservatives versus liberals. We are supposed to be above politics, intently focused on finding the truth, understanding the natural world, and applying our discoveries to improve people's lives. However, the loss of life to politicized anti-science aggression is a circumstance that we, as physician-scientists, cannot ignore. We must do our best to uncouple the anti-science from political discourse in order to save lives.

This book honors the Americans who lost their lives to anti-science aggression. It also pays tribute to the millions of Americans fated to suffer the long-term and debilitating effects of long COVID and the many children who will suffer because they lost parents or caregivers during the pandemic. In it, I warn that more deaths and disabilities will follow across the globe when children are denied access to their routine immunizations. Our entire vaccine ecosystem and programs devoted to global public health and pandemic preparedness are in trouble. With biomedical science more broadly under threat, as well as the scientists themselves, I believe that now is the time to take action to ensure they are protected.

I want to thank Johns Hopkins University Press and my editor, Robin Coleman, for his unwavering support, enthusiasm, help, and advice. I also thank Beth Gianfagna for her copyediting and assistance. This is my fourth book with Hopkins Press. As always, thank you to Nathaniel Wolf, who helps and advises me on so many editorial matters. I also want to thank Douglas Soriano Osejo and Nathan Harrington for their administrative assistance. My bosses at Texas Children's Hospital and Baylor College of Medicine, Mark Wallace and Dr. Paul Klotman, respectively, remain stalwart supporters despite a challenging political environment in Texas. For that, I am deeply grateful. Many thanks to the important administrative leaders of both institutions. At Texas Children's Hospital, they include Dan Diprisco, Lance Lightfoot, Weldon Gage, John Scales, Tabitha Rice, Michelle Riley-Brown, Angela McPike, Kelley Carville, Johnna Carlson, and Aisha Jamal. At Baylor College of Medicine, they are (among others) Robert Corrigan, Lorie Tabak, Claire Bassett, Dr. Mary Dickinson, Joe Doty, Kimberly David, Dr. Bert O'Malley, Lori Williams, Herb Butrum, Stephanie Young, Dr. Ashok Balasubramanyam, and Dr. Jim McDeavitt. A special thank-you to my new Department of Pediatrics Chair at Baylor College of Medicine, Dr. Lara Shekerdemian, and Dr. Huda Zoghbi, Professor and Chief of Pediatric Research, as well as Howard Hughes Investigator and Director of the Jan and Dan Duncan Neurological Research Institute at Texas Children's Hospital, for their support. Thank you to Dr. Maria Elena Bottazzi, my science partner for the past 20 years and codirector with me of the Texas Children's Hospital Center for Vaccine Development, and our scientists at the vaccine center and our National School of Tropical Medicine at Baylor College of Medicine. A special thank-you to Rep. Lizzie Fletcher (D-TX, and her staff), who was so kind and thoughtful to nominate both Maria Elena and me for the Nobel Peace Prize. A thank-you also to Rep. Michael McCaul (R-TX) and Rep. Joaquin Castro (D-TX), who helped to champion our Texas Children's Hospital recombinant protein vaccine along with 12 other members of the US Congress. Rep. Sheila Jackson Lee (D-TX) has been an important mentor since I first moved to Texas, as has Houston Mayor Sylvester Turner and Harris County Judge Lina Hidalgo.

I wish to thank Dr. Saad Omer, director of the Yale Institute for Global Health, for cochairing with me the Lancet Commission on Vaccine Refusal, Acceptance, and Demand in the United States of America, and for our very helpful discussions on these issues, along with the other commissioners. They include Drs. Regina Benjamin and Noel T. Brewer, University of North Carolina; Alison M. Buttenheim, University of Pennsylvania; Timothy Callaghan, Boston University; Arthur Caplan, NYU School of Medicine; Richard M. Carpiano, University of California, Riverside; Chelsea Clinton, Columbia University; Renée DiResta, Stanford Internet Observatory; Lisa C. Flowers, Emory University; Alison Galvani, Yale School of Public Health; Rekha Lakshmanan, The Immunization Partnership; Yvonne A. Maldonado, Stanford University; Michelle Mello, Stanford University; Douglas J. Opel, University of Washington School of Medicine; Dorit R. Reiss, University of California Hastings College of Law; Daniel A. Salmon, Institute for Vaccine Safety, Johns Hopkins Bloomberg School of Public Health; Jason L. Schwartz, Yale School of Public Health; and Joshua Sharfstein, Johns Hopkins

Bloomberg School of Public Health. I also want to thank the COVID-19 Vaccine and Therapeutics Task Force of the Lancet Commission for COVID-19, which I cochair with Dr. Bottazzi. They include Drs. David Kaslow, PATH; Jerome Kim, International Vaccine Institute; Heidi Larson, London School of Hygiene and Tropical Medicine; Bhavna Lall, University of Houston; Carolina Batista, MSF/Doctors Without Borders; Önder Ergönül, Koç University in Turkey; Peter Figueroa, University of the West Indies in Jamaica; Dame Sarah Gilbert, Oxford University; Mayda Gürsel, Middle East Technical University in Turkey; Mazen Hassanain, King Saud University; Gagandeep Kang, Christian Medical College in India; Denise Naniche, IS Global and the University of Barcelona in Spain, Timothy Sheahan, University of North Carolina Chapel Hill; Shmuel Shoham, Johns Hopkins University School of Medicine; Nathalie Strub-Wourgaft, Drugs for Neglected Diseases Initiative; Samba Sow, University of Maryland; Annelies Wilder-Smith, London School of Hygiene and Tropical Medicine; Yanis Ben Amor, Columbia University; and Prashant Yadav, Center for Global Development and Harvard Medical School. Thanks to Dr. Richard Horton for establishing both commissions and his vision and leadership at *The Lancet*.

Many thanks to Dr. Ali Mokdad at the Institute for Health Metrics and Evaluation, University of Washington; Charles Gaba, ACA Signups (his analysis was central to understanding COVID-19 vaccinations and deaths along a partisan divide); and Dr. Angela Rasmussen at the Vaccine and Infectious Disease Organization of the University of Saskatchewan for their sharp analyses. Also, to Drs. Eric Ding and Jorge Caballero, for pointing me to data trends on the COVID-19 pandemic both in the United States and globally; Drs. David Gorski (Wayne State University) and Jonathan Howard for their inspirational blogs and other writings; and Drs. Georges Benjamin (American Public Health Association) and Gavin Yamey (Duke University), for their observations and general support. Dr. Ruth Ben-Ghiat from New York University, Molly Jong-Fast, Anthony Scaramucci, Olivia Troye, Andy Slavitt, and Michael Moore provided essential political perspectives and expertise. I am also deeply grateful to

Dr. Michael Osterholm (director of the Center for Infectious Disease Research and Policy at the University of Minnesota) for arranging our weekly "kitchen cabinet" meetings during the pandemic to discuss a wide range of issues and bringing together extraordinary friends and colleagues. They include Drs. Ruth Berkelman (formerly at CDC and Emory University), Bruce Gellin (Rockefeller Foundation), Stephen Hahn and Margaret Hamburg (both former FDA commissioners), Penny Heaton (formerly at the Gates Foundation and now at Johnson & Johnson), and Eric Topol (Scripps Research Translational Institute).

Special thanks to Sir Alimuddin Zumla from University College London, Dr. Patrick Soon-Shiong at ImmunityBio and the *Los Angeles Times*, Dr. David Broniatowski at George Washington University, Todd Unger and Sara Berg at the American Medical Association (AMA), the National Organization for Rare Disorders (NORD), the Association of American Medical Colleges (AAMC), the American Society of Tropical Medicine and Hygiene (ASTMH), the American College of Physicians (ACP), the American Association for the Advancement of Science (AAAS), Maque García, Otis Rolley at the Rockefeller Foundation, the World Vaccine Congress, the Korean RIGHT Fund, the US-Israel Binational Science Foundation, the Immunization Partnership, the Texas Medical Association, and the Texas Pediatric Society. Also, thanks to Dr. Julie Boom at Texas Children's Hospital and Dr. Kirstin Matthews at the Baker Institute; Gabriella Stern, Stewart Simonson, and Peter Singer at the World Health Organization (WHO); and Imran Ahmed at the Center for Countering Digital Hate (CCDH) for helping to crystallize some of my thinking about the global anti-vaccine movement. I wish to thank Isabel Valdez, PA at Baylor College of Medicine and the Baylor Clinic for taking care of me when I had breakthrough COVID-19 in the BA.2.12 wave in 2022. I also wish to thank a group of physicians and scientists based currently or formerly in Atlanta, including Drs. Mark Rosenberg and Bill Foege, formerly of the CDC and Task Force for Global Health, and Drs. Roger Glass (who heads the NIH Fogarty International Center), Barbara Stoll, Venkat Narayan, Carlos del Rio, and Sudhir Kasturi. I thank Drs.

Preface

Gerald Keusch at Boston University; Peter Daszak at the EcoHealth Alliance; David Morens at NIAID-NIH, Jeff Sturchio, and Richard Roberts, Nobel laureate at New England Biolabs, for our discussions about antiscience aggression, as well as Dr. Michael Mann from the University of Pennsylvania for his willingness to share his experiences of going up against climate denialism. A special thank-you as well to Dr. Victor Dzau, president of the US National Academy of Medicine, Drs. Anthony Fauci and Francis Collins from the US National Institutes of Health, and Dr. Tedros Adhanom Ghebreyesus, director general of the WHO.

I want to express my gratitude to the organizations, foundations, and individuals who supported the research and development for our neglected tropical disease and coronavirus vaccine development program. I wish to extend my appreciation to the Houston Police Department, Texas Medical Center Police Department, Texas Children's Hospital and Baylor College of Medicine Security, and the Anti-Defamation League Southwest for keeping my family and me safe during this time of aggression. I appreciate the support of CNN, MSNBC, NPR, PBS, Comedy Central (*The Daily Show*), and Univision, as well as the major networks—CBS, NBC, and ABC—for their willingness to allow me to have a discussion with the American people about the rise of anti-vaccine and anti-science aggression during the COVID-19 pandemic. Similarly, I thank the CBC, Global, and CTV in Canada, the BBC and Sky News in the United Kingdom, and the ABC in Australia. I want to thank my family, including my wife of 35 years, Ann Hotez, who has endured the many hardships resulting from my going up against anti-science aggression, and my adult children and their spouses or significant others: Matthew and Dr. Brooke Hotez, Dr. Emily Hotez and Yan Slavinskiy, Rachel Hotez, and Daniel Hotez and Alexandra Pfeiffer. I also thank my brother, Dr. Lawrence Hotes, and his wife, Linda; my sister Elizabeth Kirshenbaum and her husband, Warren; Andi Hotes and all of her children—my nieces and nephews—as well as my mother-in-law, Marcia Frifield, together with Julia Frifield and David Frifield. Finally, thank you to my wonderful neighbors for looking out for my family.

THE DEADLY RISE OF ANTI-SCIENCE

Sent: Sunday, May 9, 2021 6:37 PM

To: Hotez, Peter Jay

dr (I use that term loosly) hotez, you print one more paper or blog suggesting the round up of free citizens to get an unproven, untested shot that is no more a vaccine than an m and m is. I will suggest to all my patriot friends that you be hunted and . . .

Sent: Sunday, May 16, 2021 8:37 AM

To: Hotez, Peter Jay

Justice on the way for you scumbag!!

1 | An Army of Patriots Turns against the Scientists

Ugly e-mails in my inbox or threats on my Twitter social media account can appear on almost any day, but for some reason the worst and most intimidating ones seem to arrive on Sundays. I remember this because Sundays are special for me. While I typically work constantly to keep up with lab meetings, my lectures, or revising the scientific papers and grant applications from our vaccine research group, on Sundays I try to do something a bit different. Unless there is some urgent business in the lab or at the college, I focus my energies on larger societal issues related to the complicated geopolitics of modern biomedical science or an expanded role for biomedicine in addressing global health inequities. My Sunday writing topics: vaccine diplomacy—the role of vaccine science collaborations in promoting international relations; combating pandemic threats through global vaccinations; vaccines and other interventions for neglected tropical diseases linked to human poverty; and, of course, combating anti-vaccine activism.

Heading a scientific research group is a demanding job, so I especially treasure Sundays, when I can step back to reflect on these big-picture concepts. I love explaining how scientific endeavor gives us unique and special powers for achieving humanitarian goals. This is the reason I became a physician-scientist in the first place: to use scientific knowledge to make the world better. Forty years ago, I obtained my PhD and MD from Rockefeller University and Weill Cornell Medical College, respectively. The motto of the former is *scientia pro bono humani generis*—knowledge for the benefit of humankind.

To receive dark e-mails or tweets on a Sunday that ominously warn of patriots hunting me down or of impending "justice" is hurtful and jarring. Hurtful, because when I decided back when I was an adolescent that I would one day become a scientist, I never imagined a segment of society turning against me or my scientific colleagues. It is still almost unbelievable how many Americans now view us as enemies. Jarring, because the threats shake me out of my usual contemplative Sundays. The e-mails are also scary—often it is not clear at first glance if the threats are real and could lead to physical harm or whether they simply reflect the anger or frustration of a sad or unhappy, solitary, or lonely individual. Nowadays, I have a system in place to alert the proper authorities in law enforcement or groups that combat hateful speech and rhetoric, such as the Anti-Defamation League (since it is generally well known that I'm Jewish), to follow up with these individuals and alert me if I or my family is under imminent threat. I will have more to say later about my frustrations when I'm under attack or the toll the attacks take on one's mental health, but one thing is for sure: The aggression manifested through e-mails, tweets, phone calls to my office, and even sometimes actual physical confrontations is relatively new and accelerating at a fast clip. Increasingly, it looks as though we now face a new normal for what it means to be a biomedical scientist in America in the 2020s.

The Big Picture: Anti-science as the Scary New Normal across Conservative America

This is a dark and tragic story of how a significant segment of the population of the United States suddenly, defiantly, and without precedent turned against biomedical science and scientists. I detail how anti-science became a dominant force in the United States, resulting in the deaths of thousands of Americans in 2021 and into 2022, and why this situation presents a national emergency. I explain why anti-science aggression will not end with the COVID-19 pandemic. I believe we must counteract it now, before something irreparable happens to set the country on a course of inexorable decline.

Throughout the twentieth century, many Americans—and possibly most of the country—understood the importance of our scientific research universities and institutions and their role in preserving national security and ensuring that the United States achieved international admiration and stature. The University of Chicago and the University of California system gave us nuclear energy; the Massachusetts Institute of Technology refined radar for military use; the University of Pittsburgh gave us the injectable (Salk) polio vaccine, and Cincinnati Children's Hospital, the oral (Sabin) polio vaccine; Boston Children's Hospital and Harvard Medical School gave us the measles vaccine; Philadelphia's Wistar Institute provided advances to make possible vaccines for rubella, rabies, and rotavirus infection; and Rockefeller and Yale Universities gave us insights into how the immune system functions in order to design new and better vaccines overall. Scientists shaped our nation's destiny through a successful Manhattan Project to defeat fascism in World War II. They provided the tools to win the Cold War against the Soviet Union and charted a path to defeat cancer and heart disease. Through advances in biomedical science, we can envision a time when we might live without fear of global catastrophic illnesses such as HIV/AIDS, tuberculosis, malaria, or neglected tropical diseases.

Over these decades, Americans understood why becoming a scientist was a special and noble pursuit. Science dedicated to the betterment of humanity ranked among our nation's highest professional callings. I caught that spirit and embarked on MD-PhD training in New York City to become a vaccine scientist working to counter neglected tropical diseases and coronavirus infections.

Now, in the years immediately preceding and during the COVID-19 pandemic, something abhorrent has taken shape to reverse this course and cause millions of Americans to distrust biomedical science and at times to view prominent US scientists as public enemies. Across the conservative strongholds of Texas and the southern United States, where I live and work, as well as in Appalachia and areas of the Mountain West, open contempt of science has become a new normal. Millions of Americans who live in conservative or so-called red states now think of scientists as dangerous shills for the pharma industry or conspirators with global elites to acquire vast wealth and power. They believe this because their elected leaders tell them such falsehoods, while each weekday evening such distortions are amplified on conservative news outlets, with published social science evidence finding links between Fox News viewership and COVID-19 vaccine refusal [1]. My use of the term "conservative" here is not based primarily on traditional policies such as fiscal responsibility and limited government. Instead, it refers to a newer phenomenon linked to what some call "Middle American Radicals" or even the "New Right" living in the southern United States or in the Midwest and Mountain West. Democratic or blue states are also not immune to anti-science attitudes, but such outlooks dominate the civil societies and political landscapes in the red ones.

The consequences are shocking: as I will detail, more than 200,000 Americans needlessly lost their lives because they refused a COVID-19 vaccine and succumbed to the virus. Their lives could have been saved had they accepted the overwhelming scientific evidence for the effectiveness and safety of COVID-19 immunizations or the warnings from the community of biomedical scientists and public health experts about the dangers of remaining unvaccinated. Ultimately, this such public defiance of science became a leading killer of middle-aged and older Americans, more than gun violence, terrorism, nuclear proliferation, cyberattacks or other major societal threats [2].

Public defiance of vaccines and vaccinations occurred for two major reasons. The most obvious was widespread "exposure to misinformation" and accepting it at face value [3]. This significantly reduced COVID-19 vaccination uptake [3]. But the other essential element, and one that is seldom explained, is how simply calling this "misinformation," or some other bland descriptor, mostly fails to convey its deliberate, well-organized, and well-financed origins, as well as its political ties. A new and tragic reality had emerged: those living in conservative areas of the nation fell victim to a new and dangerous force hitherto unnamed. "Anti-science aggression" is a more appropriate term, and in this book I call out its perpetrators—members of Congress, sitting governors, conservative news outlets, the federal courts, and a cadre of contrarian intellectuals and pseudointellectuals from universities and far-right think tanks. Moreover, I will emphasize that while anti-science aggression may have accelerated during the COVID-19 pandemic, there is much more in store. Anti-science now threatens the future of childhood vaccination programs. Declining vaccination rates and a new "low-vax" future will ensure the widespread return of ancient childhood scourges such as measles or polio. In so doing, anti-vaccine activism would reverse the United Nations global Millennium and Sustainable Development Goals [2]. It could also undermine progress in all aspects of the biomedical sciences, including gene editing and systems biology, just as it has already done for climate science and efforts to reduce global warming.

Anti-science is also globalizing. Currently a dominant killing force in the United States, anti-science has also expanded northward into Canada and now pervades western Europe as well. It even extends into African nations [2]. Anti-science has historical roots that go back more than one hundred years, to when Joseph Stalin first understood its value to an authoritarian regime like Communist Russia. Discrediting science and attacking scientists is a central theme for autocrats seeking to hold power and acquire geopolitical dominance. This is a deeply troubling and profoundly sad American tragedy but one that must be unveiled in order to prevent further loss of life and to restore science as an essential component of the American fabric.

Countering anti-science aggression will require new approaches to science communication and public engagement. In some cases, we may need to discard outdated conventions that say scientists should stick to their lab meetings, grant applications, and journal publications, while maintaining political neutrality at all costs. Especially in biomedicine, we will need a new generation of scientists who are willing to defend science in the public square, and at times even to defend our colleagues from attacks by elected officials, the media, and high-profile contrarian intellectuals—including some with scientific backgrounds. We must also work with our premier scientific professional societies and academies to help them understand how biomedical science is under threat. In some ways this resembles how the fields of climate and earth sciences were attacked during the past two decades. Ultimately, the climate scientists fought back, and they now have a system in place for mounting vigorous responses to preserve and defend their expertise. We may need to create new institutions or organizations to respond to a widening assault on biomedical science, while working in parallel with the US federal government to identify legal mechanisms to dismantle an organized and well-funded anti-science ecosystem. Now that anti-science has globalized, we might consider new roles for international organizations, including the United Nations, to help in this fight. Anti-science each year is now responsible for the loss of thousands of lives, along with disabilities on an unprecedented scale. Combating this menace is no luxury or abstract discussion. It is an essential moral imperative.

The events of the past few years have been profoundly disappointing. There was a time, especially before the pandemic, when I felt we could still hope to turn back the tide of anti-vaccine sentiments and activism.

In 2017, I wrote an article titled "How the Anti-vaxxers Are Winning" to awaken Americans to the dangers of this movement [4]. Together with the scientific community I worked hard to debunk the claims that helped found the modern anti-vaccine movement—namely, their assertions that vaccines caused autism or related conditions [5]. Then in 2020 when the pandemic struck, I had some hope that ominous predictions of large-scale hospitalizations or deaths from COVID-19 would remind everyone of the importance of having a vaccine available. I was partially right—one segment of the US (and global) population was indeed desperate for a COVID-19 vaccine. However, another very large element became vehemently opposed and defiant. The deaths that resulted from vaccine defiance reached staggering numbers and proportions.

New and Tragic Beginnings

As both a pediatrician-scientist who develops vaccines and a parent of an adult daughter with autism and intellectual disabilities, I have had a front-row seat on the modern anti-vaccine movement in America. In its early years, anti-vaccine activism first grew out of a loose confederation of small, grassroots organizations that asserted vaccines were unsafe because they contained toxic ingredients and caused autism. A 1998 paper published in the *Lancet*, one of our most prestigious biomedical journals, provided the spark. It made claims that the measles-mumps-rubella (MMR) vaccine was somehow linked to autism [6]. Even though the article was eventually retracted by the journal's editors in 2010, by then a full-on movement was under way. In time, some groups acquired financing and better organization, so that by 2019 anti-vaccine groups had enlarged their remit to wage war on pharmaceutical companies through class-action lawsuits, and to sow distrust in government-funded biomedical research and even the medical profession itself.

Pediatricians and pediatric nurse practitioners across the United States began to feel they were under siege by parents who downloaded disinformation from the Internet, which was now dominated by the rhetoric of anti-vaccine groups. Quite a few parents became frightened that routine immunizations would halt or reverse their child's development and could even cause autism or intellectual disabilities. Healthcare providers struggled to keep up with the false assertions from the anti-vaccine activists. They were often made to feel inadequate by parents who insisted their child's medical professionals failed to stay up to date with the most recent literature on the subject. The truth was, the pediatricians were *au courant* with the real biomedical science, but they were not keeping up with the quackery promoted by groups and individuals who successfully figured out how to monetize the Internet by selling phony autism cures, unnecessary nutritional supplements, or anti-vaccine disinformation books [5]. Tragically, this rise in anti-vaccine activism caused significant pockets of vaccine resistance to arise across sections of the United States, especially in some urban counties in Texas [7]. Particularly worrisome was the formation of newly organized anti-vaccine political action committees, or PACs. By 2019, anti-vaccine PACs were working with state legislatures and political candidates to promote vaccine exemptions for nonmedical reasons and to expand a pool of self-identified "conscientious objector" parents who opted their children out of the routine childhood immunizations required for school entry. In several states such as Texas and Oklahoma, the reality was that parents no longer automatically agreed to vaccinate their children. The new normal was to question the legitimacy of childhood immunizations.

Some of these same anti-vaccine organizations also began to target specific ethnic groups, including the Somali immigrant community in 2017 and Orthodox Jewish communities in 2019. Anti-vaccine activists saw these groups as insular or isolated and therefore especially vulnerable to disinformation campaigns. Because measles is one of the most transmissible viral diseases, it is typically the first breakthrough infection in communities where immunization rates decline. Not unexpectedly, by 2019 there was a rise in new measles outbreaks among Orthodox Jewish

"Anti-vaccine" versus "Anti-science"

These terms can mean different things to different people. I use the term "anti-vaccine" to refer to groups or individuals making false claims about vaccines or their harmful effects. Beginning in the late 1990s and into the 2000s, the anti-vaccine movement first alleged that vaccines caused autism or intellectual disabilities, since disproven. From there, the anti-vaccine movement evolved to incorporate a range of assertions and accusations, including false claims that vaccines cause infertility, autoimmunity, or unspecified "chronic illnesses." Today, anti-vaccine groups and leaders promote the political exploitation of vaccines and vaccine mandates, claiming they constitute instruments of government control, while simultaneously dismissing their health and community benefits.

"Anti-science" is a broader term that includes efforts to undermine the mainstream views of vaccinology as well as research conclusions in other areas, such as climate science and global warming. In biomedicine, anti-science targets multiple fields, including evolutionary biology, stem cell biology, gene editing and gene therapy, vaccinology, and virology. A prominent example features unfounded claims about the origins of the COVID-19 pandemic in China. Disinformation and conspiracy theories represent major tactics of groups and individuals committed to anti-science agendas. They undermine confidence in mainstream scientific thought and practices but also in the scientists themselves. Anti-science leaders and groups employ threats and bullying tactics against prominent US scientists. Increasingly and especially in the United States, anti-science has become an important but dangerous political movement. It increasingly attracts those who harbor extremist views. In 2021, I defined it as follows [8]:

> *Anti-science is the rejection of mainstream scientific views and methods or their replacement with unproven or deliberately misleading theories, often for nefarious and political gains. It targets prominent scientists and attempts to discredit them.*

communities in New York and New Jersey and in low-vaccination pockets in the western United States, resulting in the highest number of measles cases since the nation had first eliminated (but not eradicated) that serious childhood illness in 2000 [9]. Measles was returning, and our national status as one of the first countries to eliminate measles as a public health threat was in danger of reversing. To make matters worse, in 2022 a case of polio (the first in more than 20 years) occurred in an unvaccinated young man from one of the same Orthodox Jewish communities in New York [10]. We faced a situation, not seen in several decades, in which epidemics from once-eliminated childhood infections might return annually or every other year.

By 2019, we had gathered good intelligence about the groups driving the content of the anti-vaccine disinformation. Their lies and half-truths were spreading rapidly and extensively across the Internet and other media, causing parents to reconsider vaccines. A year earlier, a new counterorganization, known as the Center for Countering Digital Hate (CCDH), formed to expose how anti-vaccine individuals or their groups were damaging societies through their Internet presence. I had grown to admire CCDH's charismatic leader, former British Labour political adviser, Imran Ahmed. It was also not lost on me how truly awful it was that we actually needed an organization dedicated to opposing digital hate, but that was our new reality.

Ahmed and his CCDH colleagues conducted careful analyses of how disinformation spreads on the Internet. In some cases, they used sophisticated computer algorithms and other modern digital methodologies to conclude that about a dozen sources were responsible for approximately two-thirds of the anti-vaccine misinformation and disinformation on the Internet and social media. Appropriately enough, they were labeled the "disinformation dozen" [11, 12].

I remember when I first saw the list of 12 generated by CCDH. Although its findings were published in 2021, many of the dozen had been around for years, and most were already quite familiar to me. I considered these or related anti-vaccine activists or groups dangerous because they were causing baseless fears about vaccines and frightening parents. As a result, parents began denying their children access to lifesaving vaccinations in record numbers. In this way, the anti-vaccine activists helped to bring back childhood scourges we had stopped worrying about because we thought these diseases had long vanished. Exactly why they targeted parents and vaccinations remains unclear, although some news outlets report considerable financial gains for the social media companies or for specific anti-vaccine groups and individuals [13–16]. In some ways, seeing the anti-vaccine groups branded as the "disinformation dozen" was reassuring. For the first time, I had some external and unsolicited validation of what I had been saying to colleagues for some time.

For years, the anti-vaccine activists had taken aim at me personally, or my family. Their leaders publicly accused me of being a "pharma shill," claiming that I defended vaccines for financial gain, when in fact I took no money from the pharmaceutical industry. The reality was quite the opposite. I have spent my entire life developing new vaccines for poverty-related neglected diseases, also known as the neglected tropical diseases [17]. In fact, our group, known as the Texas Children's Hospital Center for Vaccine Development, had even developed a unique and patent-free, low-cost COVID-19 vaccine technology, which was transferred to local vaccine manufacturers who produced tens of millions of doses of vaccines. Those vaccines were then released for emergency use authorization in India at the end of 2021 [18], and later in Indonesia in 2022 [19]. The CORBEVAX vaccine produced and tested by Biological E. in India has been administered to 75 million children in that country. It was also approved in Botswana for the African continent. In Indonesia, the IndoVac vaccine produced and tested by Bio Farma is now one of the first halal COVID-19 vaccines for low- and middle-income Muslim-majority countries [20]. However, the disinformation dozen dismissed these realities as nothing more than inconvenient truths. Possibly their concern was that I was exposing their Internet and other media activities as disinformation.

After I had written the 2018 book *Vaccines Did Not Cause Rachel's Autism* about my youngest daughter [5], the level of aggression increased even further in the form of hate e-mails or taunts on social media, especially through my Twitter account. One of the leaders of the disinformation dozen even publicly labeled me the "OG Villain." I had to look it up. It meant I was "the original gangster"—all because I defended vaccines, explained why they did not cause autism, and how the processes leading to autism began during pregnancy through the action of specific genes and epigenetic events. I had even explained how Rachel's autism gene was discovered through whole exome genomic sequencing, using methods developed at Baylor College of Medicine's Department of Molecular and Human Genetics, the same institution where I am a professor. Her autism gene encodes a specific component of the neuronal cytoskeleton required for communication between cells in the brain [21]. This finding made sense given the pervasive behavioral and neurologic manifestations of autism. In fact, the older name for autism in the *Diagnostic and Statistical Manual of Mental Disorders*, considered by many a gold standard for psychiatrists and other health professionals, was "pervasive developmental disorder." It never made sense to me why or how a vaccine could produce such global changes in the brain, but an autism gene involved in neuronal communication made total sense.

After the *Rachel* book, harassers eventually went beyond threats on the Internet and progressed to some very unpleasant and even menacing face-to-face confrontations. In 2019, I was accosted in Houston before speaking at a public health briefing sponsored by a member of the Texas congressional delegation. Then, at a medical vaccine conference held in New York City, two individuals stalked and taunted me for defending vaccines. The following day, a large group of anti-vaccine protestors surrounded the main hotel entrance as I spoke inside. It was ironic that a decade earlier I had spoken at this same hotel about efforts to provide mass treatments for the world's neglected tropical diseases, sharing the stage with President Bill Clinton at a meeting of the Clinton

Global Initiative [22]. Now instead of receiving public adulation, my work in biomedical sciences was an object of derision or vilification by extremist groups. I found this extremely demoralizing, given that I had devoted my life's work to science for the benefit of humanity.

My world was far from unraveling, but this new aggression made me feel very unsettled and upset. I was helping to lead a group making important discoveries, including new vaccines that would save thousands of lives; our vaccine center and National School of Tropical Medicine was training a new generation of physicians and scientists who would make similar discoveries; I had also advocated for a new package of essential medicines that ultimately would reach more than one billion people affected annually by neglected tropical diseases—and yet these disinformation groups peddled a false narrative that I was somehow an enemy of the people. As I read about the history of the Soviet Union and its oppression of scientists and intellectuals—beginning with Nikolai Vavilov, the renowned geneticist who perished in a gulag, and later Andrei Sakharov, the physicist who developed the Soviet hydrogen bomb but later became an activist for disarmament and human rights, whom the authorities prevented from leaving his exile in Gorki in order to receive his Nobel Peace Prize—I began to wonder if the attacks against me represented the beginnings of something just as ominous.

Pivoting from Autism to "Health Freedom" Propaganda and Politics

As bad as things had gotten by 2019 as a result of anti-vaccine campaigns that helped fuel the return of measles in the United States and that made daily life uncomfortable for me and other scientists, nothing quite prepared us for what was about to unfold with the emergence of COVID-19. For a few weeks at the pandemic's outset, I thought there was a remote possibility that with the world worried about a looming public health impact, the anti-vaccine movement might perhaps go into hibernation or even retreat. Dr. Nancy Messonier, the director of the National Center for Immunization and Respiratory Diseases at the US Centers for Disease Control and Prevention (CDC), warned in February 2020 about serious "disruptions to daily life" [23], and I thought this might prompt many Americans to clamor for safe and effective COVID-19 vaccines.

Many Americans were indeed eager to receive their COVID-19 vaccinations. However, the opposite also happened. As the pandemic unfolded, it became essential to implement new social-distancing measures and face mask requirements as the government launched its Operation Warp Speed program to accelerate the development of new mRNA (and other technologies) COVID-19 vaccines. These activities mobilized a counteroffensive in the form of a new version of the anti-vaccine movement that cared less about autism and more about personal liberties. Anti-vaccine activists embarked on a new propaganda campaign of "health freedom" or "medical freedom." By invoking freedom rather than autism, the anti-vaccine movement had become increasingly political, first linking to the Republican Tea Party in Texas before expanding to far-right and conservative groups across the country [24]. This trend had first begun in the 2010s, as anti-vaccine groups in America and extremist politics on the right found each other in unique and interesting ways.

First, the anti-vaccine movement needed to reenergize, in part because its foundation based on phony autism links was unraveling. Many of us in the biomedical community provided convincing evidence that vaccines did not and could not cause autism. Detailed and summarized in my *Rachel* book, the evidence included large epidemiological studies showing that kids who received vaccines were no more likely to acquire autism than unvaccinated kids. Such findings were further supported by laboratory animal studies that could not find a relationship between vaccines and neurologic changes [5]. In addition, we offered a compelling alternative narrative backed by years of biomedical research supported by the National Institutes of Health, the Simons Foundation,

Autism Science Foundation, JPB Foundation, and other philanthropies, which identified multiple autism genes expressed in early fetal development. Therefore, while the thread of autism and vaccines continues among anti-vaccine groups, it is no longer the major driver. The bottom line was that anti-vaccine groups needed a new angle or message to maintain cohesion and to receive an infusion of funding. Linking to far-right elements provided both, even assisting them in the formation of new PACs to lobby or educate state legislatures voting on school vaccine exemptions. One of the first such PACs was Texans for Vaccine Choice in 2015, and I believe such politicization by creating PACs and related mechanisms benchmarks a period when children increasingly attended school without their full complement of routine immunizations ordinarily required for attendance [24].

At the same time, the political far-right sought more adherents and followers. Anti-vaccine groups, including the disinformation dozen, provided "one-stop shopping" to acquire a significant cohort of people who might follow in the path of extremist groups. Through the anti-vaccine movement, potentially thousands of new people could augment the ranks of those professing allegiance to an extremist agenda. In this way, the anti-vaccine groups and the far-right found and mutually reinforced each other. Even groups linked to the January 6 insurrection at the US Capitol, such as the Proud Boys, began promoting an anti-vaccine agenda. In turn, anti-vaccine rallies could count on the insurrectionist groups to join their protests [25].

The coalescing of the anti-vaccine and far-right activities created new synergies to amplify a reinvigorated anti-science disinformation empire. Its elements included open expressions of distrust of science by ultraconservative members of the US Senate, House of Representatives (especially its House Freedom Caucus), and some governors, and even efforts by these elected officials to humiliate scientists in public or threaten their employment. These will be discussed in more detail, but they included introducing bills to demand the firing of a prominent government biomedical scientist who served the nation for decades, holding hearings or roundtables to promote anti-vaccine viewpoints, and harassment of nongovernment biomedical scientists who work at universities or nonprofit organizations. Their actions against science and scientists operated through unprovoked and unjustified public pronouncements and accusations—and much more.

High-profile conservative news outlets such as Fox News regularly amplified these messages [1], and the scientific community came under fire from the networks' highly viewed nighttime anchors. A third group, a carefully cultivated cadre of intellectuals or pseudointellectuals from universities and think tanks, gave academic cover to the assault from the far-right by producing well-crafted false narratives about the harmful effects of COVID-19 preventive measures. In turn, these activities fueled hate from White nationalists and other extremist groups, along with unwell or uneducated individuals, who interpreted the public commentaries on news outlets and podcasts as dog whistles. Finally, even the federal courts had an indirect role, especially conservative justices, who reversed government mandates for vaccines, masks, and other precautions, although in some cases it became difficult to attribute their rulings to ideological leanings versus their bona fide interpretation of the law.

The impact of such anti-science activities was immediate on two fronts: First, it made scientists fearful of speaking out to defend the science of COVID-19 vaccines and preventions or to come to the aid of our colleagues under attack. The same might be said of many of our scientific and professional societies. More important, it caused populations in conservative stronghold areas to shun or defy COVID-19 vaccines and other preventions. This led to massive losses in human life as the virus ripped through unvaccinated populations across the nation, but especially in these regions.

The attacks against individual prominent US scientists were especially rough. Among the most ominous were direct threats from individuals with White supremacist leanings. In some cases, they would refer to themselves as "patriots" fighting for health freedom. In the spring of 2021, several e-mails I received stand out, including one that claimed "armed white nationalists aren't going to take it that death shot. Give to the blacks and Hispanics and SCREW OFF from the rest of us." Another adds, "will suggest to all of my patriot friends that you be hunted." My first response was one of indignation. After all, our nation is one built partly on our great research institutes and universities. Scientists helped America achieve victories in World War II and the Cold War, among many other accomplishments. My father, Eddie Hotez, who fought in the Pacific Theater—Okinawa, Saipan, Philippines—as a 19-year-old US navy ensign, was proud that I became a scientist. He was buried in 2015 in a Jewish cemetery in Avon, Connecticut, with full military honors. My father was a real patriot, and someone who made a point to hoist the American flag at our home every Memorial Day. He told me, my two brothers, and sister what it was like to serve on a Landing Ship Transport vessel during maritime landings and amphibious assaults, and as kids we would flip through his book of old photographs from when he was in the navy. I learned from my dad what it means to serve our nation. But Eddie also understood that there were other ways to serve. I remember how excited he was when I would testify before Congress, which I have done on multiple occasions over the years, especially regarding the government's commitment to global health, or about the concepts of vaccine diplomacy. When I became microbiology chair at George Washington University from 2000 to 2011 (before relocating to Texas) I was a frequent visitor to the House or Senate office buildings or the Eisenhower Executive Office Building of the White House. My dad loved to hear what happened in my meetings. Later, I had the privilege of serving as President Barack Obama's US science envoy for the Middle East and North Africa, where I helped to establish new vaccine diplomacy initiatives with research institutes and universities. As an American scientist, I felt that I was the patriot, not the antivaxxers marching with the Proud Boys.

Many of the threats contained Nazi or Holocaust imagery and made it clear that the authors knew I was a Jewish scientist. Anti-vaccine groups began embracing the same anti-Semitic beliefs and sentiments as White nationalist groups. In one of my first in-person public speeches during the COVID-19 pandemic, which I gave in a Reform synagogue in Houston, Texas, during the Jewish high holidays in 2021, I was heckled and stalked by two individuals who made it past security. I could not hear most of their shouts or taunts, but I remember they included at least one pharma-shill accusation and a claim about the dangers of mRNA vaccines. Fortunately, they were escorted out by security without protest, but it was clear that anti-vaccine activists were prepared to demonstrate their disrespect for a Jewish house of worship, and that far-right extremism and anti-Semitism had woven themselves into the fabric of the movement's latest version.

As the pandemic progressed in 2021 and 2022, the public attacks against me and other US scientists became more prominent and widely viewed. You might think that these attacks would come only from the fringes or the extremists. However, in June 2021, the governor of Florida sought to discredit me during a highly watched nighttime Fox News interview together with the anchor, Laura Ingraham [26]. This happened again with Rep. Marjorie Taylor Greene (R-Georgia) on Steve Bannon's *War Room* podcast. On a later Fox News broadcast in February 2022, Tucker Carlson accused me of being a "charlatan" on the same day that I was co-nominated (with my more than 20-year science partner, Dr. Maria Elena Bottazzi) for the Nobel Peace Prize for our work to develop and produce a low-cost "people's vaccine" to prevent COVID-19 globally.

Predictably, these episodes triggered the anonymous threats by e-mail and on social media, as well as the direct physical confrontations. However, I was not the only one who experienced such attacks from ultraconservative elected officials and conservative news sources. Several other scientists working to halt the COVID-19 pandemic, most notably Dr. Anthony Fauci, the director of the National Institute of Allergy and Infectious Diseases of the US National Institutes of Health, and several other prominent virologists or epidemiologists, came under threat. A survey conducted by *Science* magazine found that almost 40% of

COVID-19 scientists reported attacks via e-mail, social media, or phone, or even physical confrontations [26]. Among the major consequences were workplace anxiety, depression, becoming fearful about personal safety or loss of reputation, social isolation, and decline in productivity. A study conducted by *Nature* magazine reported similar findings [27]. The attacks produced public health consequences. Many scientists under threat began to avoid speaking out in public or to reduce their overall public engagement [26]. This created a vacuum now filled by those who sought to question the severity of the COVID-19 pandemic and twist the facts in such a way as to undermine the benefits of COVID-19 prevention measures, including vaccines of course, but ultimately also masks and social distancing mandates.

Far-right and conservative assaults on science and scientists are not necessarily new. In his 2006 book, *The Republican War on Science*, the journalist Chris Mooney outlines the rise of anti-science activities that first began with the 1964 presidential campaign of Sen. Barry Goldwater (R-Arizona) [28]. Anti-science activities later accelerated in the Nixon administration and expanded even further when Rep. Newt Gingrich (R-Georgia) served as Speaker of the House during the 1990s. In the 2000s, prominent climate scientists came under attack.

While it can take a long time to see the negative effects of undermining climate science, the assaults on biomedical science were different because they led to an immediate loss of human life—at a devastating level. The most game-changing consequence of the attacks on biomedical science and scientists during the COVID-19 pandemic was the concerted effort by elected officials, conservative news outlets, and public intellectuals from the far-right to discredit COVID-19 vaccines. These destructive activities created a program of propaganda to make it appear that these vaccines are highly unsafe or that they do not actually even work. Moreover, they hammered on these points, night after night on cable news, during the day in the halls of Congress and red-state capitals, and among gatherings of conservatives who aligned with the extremists. The contrarian intellectuals piled on through commentaries and podcasts, thereby providing a veneer of legitimacy for this entire operation. In so doing, a new anti-vaccine ecosystem based on far-right leanings and teachings evolved to promote a culture of vaccine defiance and refusal. This became especially apparent after May 1, 2021, the date by which anyone in the United States who wanted to get a COVID-19 vaccine could do so. Despite the widespread availability of inoculation from May 1 until the end of 2021, I estimate that approximately 200,000 unvaccinated Americans needlessly lost their lives to the virus. I show how that number was derived in chapter 3.

In-depth analyses from the *New York Times* [29–31], the political and health analyst Charles Gaba and National Public Radio [32–35], and *Axios* [36] show unambiguously that most of these 200,000 deaths occurred in Republican stronghold (red) states. Partisan leanings were strongly associated with the likelihood both to be unvaccinated and to lose one's life to COVID-19. The "redder" the county in terms of the percentage of Republicans, the higher the loss of life. Vaccine hesitancy and refusal emerged as a new form of allegiance to a canon of health freedom propaganda points put forward by members of the US Senate and House Freedom Caucus, local elected officials, Fox News [1, 37] and other news outlets, and far-right podcasters and radio announcers. Many Americans paid with their lives for this sense of belonging. They perished as victims of anti-vaccine aggression.

The Other Big Picture: This Won't End with COVID-19

This is the story of how the anti-vaccine movement became a political movement promoted and endorsed by elected officials in the highest reaches of the federal and state governments in the United States. It reports how their statements, rhetoric, and actions were amplified through the most highly viewed television news programming in America. It explains why anti-science now kills more Americans than global terrorism, or other deadly societal forces and social determinants [2, 38] but also how we might mount a counteroffensive to save American lives. To do so, I will explore the possible motives for anti-vaccine and anti-science aggression and how it shifted from efforts to monetize the Internet by promoting products claiming to prevent or treat autism to a political agenda and a means to assert control and demand allegiance to authoritarianism, especially on the far-right.

It's not just the science. American biomedical scientists are also under attack, and I will explain why this situation has come to resemble past attempts from authoritarian regimes in the twentieth and twenty-first centuries to portray scientists as public enemies. While these aspects of the antagonism have become especially obvious during the 2019–22 pandemic, I predict it will not end with COVID-19. The authoritarian embrace of anti-science is not new at all; rather, it represents a well-established practice that began with "the great purge" in Stalin's Communist Russia and continues to the present through right-wing groups in the United States but also in Canada, Latin America, and Europe. This is a full-on globalized initiative. I will explore the writings and thoughts of experts investigating the inner workings of authoritarian governments, including political scientists and prominent commentators. Although I spotlight the attacks against biomedical science and scientists in America, I will also draw parallels to the attacks on predictions of climate change and how prominent climate scientists have faced threats that resemble the ones now experienced by the biomedical science community.

The consequences of allowing this political assault on science to continue unopposed are unacceptable. Not stopping at COVID-19 vaccines, it will eventually spill over to routine childhood vaccinations and other aspects of pediatric practice. During the pandemic there were significant declines in childhood vaccinations as a result of social disruptions that interfered with well-child visits to the pediatrician [39–42]. The concern is that immunization rates will not return to their original levels because of widespread anti-vaccine activities [43]. For example, one general practitioner in the United Kingdom noted about the MMR vaccine, "This is anecdotal, but we're finding quite a lot of parents saying they have researched the vaccine and are refusing it.... Covid vaccine hesitancy seems to have impacted on it, unfortunately. There seems to be a loss of trust, which is both sad and worrying" [44]. Along similar lines, those opposed to COVID-19 prevention measures in the United States falsely claim that new changes in CDC developmental milestone guidelines (made during the pandemic) reflect unprecedented developmental delays resulting from widespread mask-wearing by parents. Some anti-vaccine groups may even aspire to convince parents to skip routine pediatrician visits, because these represent the times when most infants and toddlers receive their vaccines [45].

If the momentum that an anti-vaccine movement built around spurious freedom claims during the COVID-19 pandemic continues to accelerate and extend across the pediatric vaccine ecosystem, immunization rates will decline precipitously. This could bring back childhood illness on a massive scale and reverse hard-won gains in global health. According to the Global Burden of Disease Study 2019 from the University of Washington's Institute for Health Metrics and Evaluation, between 2000 and 2019 (the year prior to the pandemic) we had achieved up to an almost 80–90% decline in under-five-year-old childhood deaths from vaccine-preventable diseases such as measles, pertussis (whooping cough), tetanus, diphtheria, and meningitis [46]. Much of that progress can be attributed a new global infrastructure to support vaccines and vaccinations that included Gavi, the Vaccine Alliance, together with UNICEF, the World Health Organization, the Gates Foundation, various pharmaceutical companies, and civil society [47]. Now an emboldened anti-vaccine movement works to put this infrastructure in jeopardy. Measles is accelerating, and more than a dozen African nations are at high risk for a return of polio [48]. As noted above, the United States, the United Kingdom, and some European nations are also at risk [39–45, 49–51]. Our entire vaccine ecosystem has become extremely fragile. Beyond its effect on vaccinations, anti-science could soon contaminate all aspects of biomedical science and erode many recent impressive discoveries in gene editing, systems biology, nanomedicine, stem cells and organoids, neural networks, or functional brain imaging. Individual scientists whose work has led to these advances could themselves come under increasing attack. America is descending into a period of darkness, with scientists portrayed as enemies.

While anti-science has become "a dominant and highly lethal force" [8], it is not too late to stem the tide. However, success on this front requires us to fully comprehend the depth and breadth of anti-science in the United States and around the world. Right now, we are at the end of a new and ugly beginning. We still have options to turn this situation around to defend and protect science and scientists, but our window for achieving this goal may soon shut.

Let me tell you how it will be.
And I don't care if you agree,
'Cause I'm the Vaxman.
Yeah, I'm the Vaxman.
If you don't like me coming 'round
Be thankful I don't hold you down ...

—Phil Valentine, deceased conservative radio host

2 | Health Freedom Propaganda in America

One of the sadder aspects of the COVID-19 pandemic occurred among those who were hospitalized after they refused to get vaccinated, including those who expressed utter disbelief that they might die from their illness. Nurses across the country reported instances in which patients refused to admit they were dying from COVID before they finally succumbed [1]. Equally heartbreaking were the patients who finally agreed to get vaccinated as their COVID-19 illness progressed, not understanding that vaccines are preventative and must be administered ahead of the illness, or those who came to realize that they had made a grievous error and pleaded with their friends and colleagues not to make the same fatal mistake.

Among the notable deaths from COVID-19 in 2021 were several conservative talk-radio hosts who demonstrated their defiance against US government health recommendations by publicly condemning vaccines and refusing to be vaccinated [2-6]. A Christian preacher from Tennessee who was syndicated nationally asked a guest if the US

COVID vaccination program "could... be another form of government control of the people?" [3]. Another reworked the 1966 Beatles song "Taxman" as "Vaxman," while a third declared that the government was "acting like Nazis" [2]. A Florida radio host (and former broadcaster on the conservative Newsmax outlet) wrote on Facebook: "Why take a vax promoted by people who lied 2u all along about masks, where the virus came from and the death toll?," although as he was dying in the hospital, he encouraged his friends to get vaccinated [4, 5]. All four of these men died in August, likely a result of the wave of a highly transmissible and lethal Delta variant of the SARS-2 coronavirus that struck the southern United States [2–5]. One journalist regretfully wrote, "All had used their platforms to push various bits of misinformation about the virus, masks and the vaccines, and all were ultimately brought low by the very virus they refused to take seriously. Sadly, several recanted of their anti-vaccine stances in particular before passing away, but by then it was too late" [5].

These COVID-19 deaths continued in 2022. In January, a "far-right podcast host" (as described by the *Daily Beast*) and vaccine critic died from the illness after "returning from a recent patriot conference," according to a statement posted on social media, believed to be an ultra-conservative "ReAwaken America Tour" held in Texas [6]. A leading QAnon influencer and promoter who claimed that COVID-19 vaccines killed people and called for the public execution of Dr. Anthony Fauci, the director of the National Institute of Allergy and Infectious Diseases and chief COVID-19 adviser to the Biden administration, also succumbed to the virus [7].

Many law enforcement officers also refused vaccinations, in some cases resigning from their jobs on the police force rather than agreeing to be immunized. Tragically, they too lost their lives on an unprecedented scale. According to the National Law Enforcement Memorial and Museum, in 2020 and 2021, COVID-19 was the leading cause of death among US law enforcement, with a 65% increase in these deaths from the first to the second year of the pandemic [8]. This meant that most of the COVID law enforcement deaths occurred after vaccines were widely available. On October 18, 2021, Washington governor Jay Inslee implemented vaccination mandates for state employees, but in a video that went viral on Twitter and other social media platforms the same day, an unnamed sergeant tearfully announced his resignation from the Washington State Patrol after 17 years of service. He stated, "Due to my personal choice to take a moral stand . . . for medical freedom and personal choice I will be signing out of service for the last time today" [9]. One of his colleagues, a 50-year-old Washington State trooper named Robert LaMay, died from the virus weeks after resigning. His sign-off ended by saying Governor Inslee "can kiss my ass" [10]. Previously, Officer LaMay stated that "the government should not be mandating vaccines" and told Seattle radio station KOMO: "You should not be forcing this. . . . If you start forcing things like this, what's next?" In October 2021, before he contracted COVID, LaMay made an appearance on Fox News and was praised by the anchors for his stance. Many in the media criticized the network for failing to update their viewers about the former trooper's subsequent death [11]. Law enforcement was not alone, as firefighters and other first responders also resisted COVID-19 vaccinations.

The common grievance voiced by these conservative talk-radio hosts, the police, and police unions was that mandated vaccines, despite their proven effectiveness against the then-circulating Delta variant of COVID-19, represented violations of American freedoms. This demanded a vigorous response, including drawing their line in the sand in the form of vaccine defiance or refusal. They did so in the name of health freedom (some called it medical freedom) and claimed this as a fundamental American principle. Their views had support from some of the highest levels of the US government. On September 30, 2021, more than a dozen conservative members of the House of Representatives, including many belonging to the House Freedom Caucus, introduced a bill—H.R. 5471, the Health Freedom for All Act, to the 117th Congress to block the secretary of labor from mandating COVID-19 vaccines [12]. Such legislative action anticipated an announcement by the Biden White House to work with the labor secretary and through OSHA (Occupational Safety and Health Administration) oversight to demand that businesses with more than 100 employees require COVID-19 immunizations.

When the White House announcement finally came in November [13], it set off an immediate backlash from conservative groups and elected officials from the GOP, including the governors. Ultimately, groups on both sides of the issue vented their extreme frustrations. The Biden White House expressed horror at the huge and needless losses of human life from those who refused vaccinations despite their widespread availability, effectiveness, and safety. In my public media appearances, including cable news channels, podcasts, and radio broadcasts, I supported the White House and described these mandates as evidence that the federal government was trying everything within its constitutional powers to increase vaccine coverage and save lives. Countering these measures were the GOP governors from more than a dozen states who almost immediately following the White House announcement filed lawsuits to halt these actions on grounds that they violated personal civil liberties [14].

This standoff reached a fever pitch on Sunday, January 23, 2022 (one year after the January 6 insurrection), when thousands of protestors marched in Washington, DC. They held a rally at the Lincoln Memorial to voice their opposition to employer or school COVID-19 vaccine mandates [15]. The events that Sunday were championed by leading anti-vaccine activists together with newly created federal employee organizations, such as Feds for Medical Freedom and DC Firefighters Bodily Autonomy Affirmation Group. The protests also included representation from White nationalist extremist groups such as the Proud Boys, linked to the January 6 insurrection and storming of the US Capitol. Prominently displayed were signs declaring, "I call the shots, not you," "my body my choice," and "medical freedom for our heroes." A bus carrying the protestors displayed a WANTED poster of Dr. Fauci, as well as Bill Gates and CDC director Dr. Rochelle Walensky, and this became a centerpiece for the rally. Beyond the presence of the Proud Boys at the anti-vaccine mandate rally, several news outlets further reported a range of connections between anti-vaccine groups and the storming of the US Capitol one year before. According to CNN, several prominent anti-vaccine activists helped to coordinate the "stop the steal" protests [16], while some of those arrested for their participation in the January 6 insurrection were well known for their anti-vaccine views and activities [17, 18].

Ultimately, the Supreme Court agreed with the Republican governors and the protestors. The 6 to 3 vote divided along a strictly partisan line: the six justices striking down the employer mandates were Republican presidential appointees, and the three justices supporting them were appointed by Democratic presidents. Matt Ford, a staff writer for the *New Republic*, a left-leaning American political magazine, excoriated the Supreme Court ruling: "The court's conservative justices went out of their way to reconfigure the OSHA mandate into something that it wasn't, then they kneecapped it for violating standards that they read into federal laws" [19]. Ford went on to detail fundamental flaws in the judgment, explaining why public health measures constitute appropriate OSHA activities, even if they are not exclusive to the workplace.

However, in April 2022, the Biden administration successfully appealed some of the lower court rulings related to vaccine mandates. Such actions set in motion efforts by federal judges on either side of the issue to wrangle over public health recommendations by the CDC and other public health agencies. This set a dangerous precedent in which courts sympathetic to health freedom plaintiffs might routinely challenge public health recommendations intended to save lives.

Origins

The health freedom face of the anti-vaccine movement and its association with American libertarianism or extreme conservatism did not begin with the COVID-19 pandemic. A closer look at the health freedom movement reveals even older roots in American history, possibly as old as the nation itself. At its core, the central tenets of health freedom emphasize protests regarding government infringement on family or personal healthcare decisions [20]. Since the founding of the American colonies and their declaration of independence from Great Britain, several historical figures stand out for their staunch views concerning health freedom. There is also a second but less intuitively obvious component of health freedom—the right to use unproven or bizarre medicines and treatment approaches. Implausible "miracle" cures are a common historical thread that runs through the health freedom story in America. Relevant to COVID-19, this might explain why hydroxychloroquine or ivermectin were touted as viable alternatives to vaccines and therefore came to play a leading role in vaccine defiance.

Dr. Benjamin Rush is a revered figure in the history of American medicine. A signer of the Declaration of Independence, the Continental Army's surgeon general, distinguished medical school and chemistry professor, and one of the founders of modern psychiatry, many consider Rush one of the great public intellectuals in colonial history. Less known was his staunch opposition to government intrusions in health and medical treatment decisions, with some arguing that he proposed medical freedom language for the US Constitution, alongside religious freedoms. Notably, Dr. Rush was a contemporary of Samuel Thompson, and the two are believed to have met [20, 21]. Thompson was a proponent of the use of botanicals for medicinal purposes and believed that such activities should not require formal medical training or be subject to licensing requirements [20, 21]. Unrestricted rights to peddle unproven treatments became front and center to health freedom, and these ideas go back to the founding of the American republic.

Medical freedom continued as an important theme during the early 1800s, before it eventually receded with the American Civil War. However, in the early twentieth century, a new umbrella organization for alternative medical practices, which included homeopaths, osteopaths, eclectic medicine (a throwback to the botanical remedies of Thompson), and some Christian Scientists, was first established under a new National League of Medical Freedom [21]. Abraham Flexner, a distinguished educator and brother of Simon Flexner (the inaugural director of the Rockefeller Institute of Medical Research), led medical education reforms and sought to discredit these alternative approaches. After an exhaustive (and presumably exhausting) tour by train and other means across the United States, in which he visited many substandard medical schools, Flexner wrote a detailed, candid, and at times scathing summary of their inadequacies [22, 23]. In his now famous report, he wrote about his visit to the Georgia College of Eclectic Medicine and Surgery in February 1909: "Nothing more disgraceful calling itself a medical school can be found anywhere" [23]. Regarding the two medical schools Flexner visited in Arkansas that year, he offered, "neither has a single redeeming feature." In contrast, he expressed far greater enthusiasm for medical schools linked to well-established universities, especially those tightly connected to active scientific departments publishing in reputable scientific journals. Flexner's report provided the academic cover that state medical or education boards required to close many substandard medical colleges committed to homeopathy or eclectic medicine. This left only those institutions that practiced scientific medicine and research, as exemplified by the newly established Johns Hopkins University School of Medicine, as the inheritors of true medical education. The new push for scientific medical education led to an important new generation of US medical schools linked to powerful academic hospitals and medical centers, such as the Texas Medical Center in Houston where I work, the nation's largest medical center. However, the organizations allied under medical freedom remained a powerful lobby for many years thereafter and readily endorsed an Anti-Compulsory Vaccination League upon its arrival in the United States from England [24]. Tensions between scientific medicine and homeopathy or eclectic medicine flared around several practices, including vaccinations, and William Osler, MD, one of the four founders of Johns Hopkins medical school, was notably vocal in his objection to the "anti-vaccinationists" [20, 21, 24].

By the middle of the twentieth century, two independent branches evolved. The first was committed to the scientific principles and underpinnings of medical practice. With the spectacular successes of Drs. Jonas Salk and Albert Sabin in developing and testing the first polio vaccines, the American people saw firsthand the immediate public health impact of a new vaccine and how it allowed them to live free from fear that their children would become paralyzed during late summers and early falls. Other vaccines quickly followed, including new antiviral vaccines for measles, mumps, rubella, and hepatitis B; and for pneumococcal and *Haemphilus influenzae* type B bacterial infections, respectively. This period also coincided with a sharp rise in public support of medical research, especially by the federal government through its funding of the National Institutes of Health.

In contrast, a parallel track that focused on unconventional cures and health freedom propaganda also emerged. During the 1950s, a National Health Federation formed as a California-based lobbying group for natural medicines; untested medical devices such as radionics, which proposed treating disease with radio waves or electromagnetic radiation; dietary supplements; and other alternative medicines. In time, the National Health Federation began to promote resistance to vaccination mandates, or as it currently states on the "mission and values" section of its website: "The right to protect ourselves and our children from unnecessary and often dangerous, childhood vaccines" [25].

Both the National Health Federation and the far-right John Birch Society, which was formed just a few years later, began to promote laetrile, a chemical discovered as a natural product in the seeds of several fruits, that subsequently was touted as an alternative cancer cure [26]. However, after clinical trials in 1977 failed to show any benefit for cancer sufferers, the US Food and Drug Administration (FDA) and the American Medical Association (AMA) worked to halt its use and to ban its interstate commerce. Such actions prompted the John Birch Society and other conservative groups to object to undue interference from the FDA and US government, very much in line with the dual nature of the American health freedom movement—protesting government interference and promoting unproven cures. These activities culminated in Rep. Ron Paul (father of Kentucky senator Rand Paul) sponsoring the Health Freedom Protection Act (H.R. 4282) in 2005–6, "to provide that a food or dietary supplement is not a drug solely because the label or labeling contains a claim to cure, mitigate, treat, or prevent disease" [27]. The act shielded many nutritional supplements from FDA oversight, thereby helping to fuel a multibillion-dollar "natural" health products industry, which decades later may have spun off financial backers for anti-vaccine groups [28].

The Rebirth of Medical and Health Freedom

In its modern form, the anti-vaccine movement gained momentum in the early 2000s after a published study alleged that the measles-mumps-rubella (MMR) vaccine replicated in the colon of some children to cause autism [29]. By the time journal editors retracted the MMR paper in 2010, anti-vaccine groups had grown in both size and funding [30]. Several leading anti-vaccine activists are now listed by the Center for Countering Digital Hate (CCDH), including those that CCDH claims constitute a "disinformation dozen," which accounts for a substantial amount of the disinformation on the Internet. Despite a massive effort by the scientific community to debunk the vaccine and autism connection, the anti-vaccine activists or groups continue to peddle this concept. A common feature is that such groups are not easily deterred by scientific evidence, so the phony autism assertions persist, although with somewhat diminished prominence. The disinformation dozen and other anti-vaccine activists or groups had to face a new reality—to maintain both their relevance and revenue they would eventually need a new spin or angle.

In many cases they found it by resurrecting health freedom. The years 2014 and 2015 benchmark this new aspect of the anti-vaccine movement. In Southern California, especially Orange County, so many parents denied their children access to MMR and other school-entry vaccines that a serious breakthrough measles epidemic ensued. In response, the California legislature introduced and passed Senate Bill 277, led by pediatrician Dr. Richard Pan, which eliminated vaccine exemptions for nonmedical reasons [31]. Closing vaccine exemptions was a surefire and direct way to ensure that children would get their lifesaving vaccines and stay protected from measles and other catastrophic pediatric infections. However, the actions of the California legislature had the unintended consequence of rejuvenating anti-vaccine activists. As we headed into the 2010s, health freedom propaganda became the newest feature of anti-vaccine activism—sort of a version 2.0. It has three central tenets [20, 32]:

1. Closing school vaccine exemptions constitutes unwarranted government intrusion into the health freedom and medical privacy of families. Vaccine mandates are unacceptable to the point of being un-American, whereas vaccine choice constitutes a necessary and essential element of core American values.
2. There is an unholy alliance between the pharmaceutical industry and state or federal governments, in some cases fueled by secretive payments or backroom deals. Health freedom activists demonize Big Pharma, believing that many vaccine scientists accept bribes or kickbacks in exchange for endorsing vaccines. They refer to such doctors as "pharma shills." These false accusations provide a rationale for groups wishing to promote unproven or spectacular cures, by claiming that the pharma industry covers up their beneficial effects to promote newer and more expensive medicines or vaccines.
3. Health freedom further promotes a belief system that relies on a pseudoscience that extols the benefits of "natural immunity" to the actual infection, which in some cases could be bolstered by nutritional supplements. In contrast, vaccines are mostly unnatural and toxic. Therefore, those reluctantly agreeing to accept a vaccination must first be offered "informed consent," including an exhaustive list of potential or real side effects.

Although it might have started in California, the health freedom movement found its full expression and home in Texas. The first PAC committed to "vaccine choice" was created in Texas in 2015 [33]. Its mission "is protecting and advancing informed consent, medical privacy, and vaccine choice through influential public policy, quality educational resources, and training to an engaged, connected community" [34]. According to Rekha Lakshmanan from the pro-vaccine organization The Immunization Partnership, based in Texas, and Jason Sabo from Frontera Strategy, "Texas legislators, particularly Republicans leery of primary election opponents, began to fear electoral consequences if they were to support vaccines or oppose anti-vaccine activists and their agenda" [33]. Lena Sun from the *Washington Post* reported that the rhetoric used by the Texas anti-vaccine lobby is "libertarian and anti-government," with strong ties to the Tea Party and Empower Texans, a very powerful conservative political organization [35, 36]. In an article published in the *Texas Observer* before the pandemic, Sarah Davis, a pro-vaccine, moderate Republican in the Texas legislature, reflected about the anti-vaccine lobby in Texas and Texans for Vaccine Choice. She said, "It's really scary how impactful they are.... Last session I couldn't even get any hearings on my pro-vaccine bills. They really have been able to bully and intimidate a lot of members. And they've made vaccines controversial, which makes members nervous.... It really shouldn't be controversial" [37].

By this time, I had relocated to Texas (from Washington, DC) to expand our research group's scientific activities and build a world-class vaccine research center and tropical medicine school at the renowned Texas Medical Center. The Texas Medical Center is the world's largest, with over 100,000 employees, 60 institutions, several medical schools (including the Baylor College of Medicine, where I became a professor and dean), and important hospitals and hospital systems, including the MD Anderson Cancer Center and the Texas Children's Hospital. Our vaccine center is based in the Feigin Center, the major research building of Texas Children's Hospital. The goal of our faculty and scientific staff is to apply the horsepower of the Texas Medical Center to solving real-world global health problems, especially as they relate to neglected tropical diseases. This was an extraordinary opportunity to apply the concept of translational medicine to making vaccines for the world's diseases of poverty. However, as anti-vaccine legislation and political activities ramped up, I became a highly visible vaccine scientist and therefore target of the Texas anti-vaccine lobby. In turn, as someone who was helping to lead vaccine research in the state of Texas, I felt an obligation to become a public defender of vaccines. In 2016, I warned about the dangers of what was happening in the state [38] and then nationally, in 2017 [39], before writing an entire book debunking the claims that vaccines could cause autism [30].

The adverse consequences of health freedom propaganda for the schoolchildren of Texas have been profound. Over the past decade, the percentage of students with one or more so-called conscientious vaccine exemptions "on file" increased approximately threefold, although for Texas private schools there was a fourfold increase [40]. Nationally, our faculty identified 15 urban counties, most located in western states, including 3 in Texas, where there were large numbers of vaccine exemptions [41]. For the most part, these areas hosted strong anti-vaccine lobbies committed to the principles of health freedom. By 2019, several had experienced measles outbreaks [42], compared with the absence of measles in areas of high vaccine coverage. Based on activities seen so far in state legislatures around the country, we might expect similar resistance to COVID-19 vaccinations for school entry [43].

O Canada: Health Freedom Globalizing

During the pandemic, anti-vaccine groups expanded their ties to conservative groups [44], as health freedom propaganda and its adherents spread out of Texas and across the nation. In 2021, they gained even greater strength, fueled by President Biden's call for employer COVID-19 vaccination mandates. Early in the fall of 2021, I wrote about how the US health freedom movement was beginning to "contaminate" Canada [45], especially in some of the more conservative western areas of the country, which were tuning in to Fox News and other US-based conservative news outlets. Soon Canadian citizens would begin launching anti-vaccine protests.

These actions culminated in a so-called Freedom Convoy of trucks that originated in different parts of Canada a day prior to the January anti-vaccine rally in Washington, DC, before converging on the Canadian capital of Ottawa a week later. Initially it started when unvaccinated American truckers were denied entry into Canada beginning January 15, 2022. Then, starting on January 22, the estimated 15% of Canadian truckers who were unvaccinated would no longer be allowed to cross the border into the United States. Through GoFundMe, GiveSendGo, and other crowd-sourced support organized by conservative leaders [46], together with vocal encouragement from Fox News and other news outlets, several thousand protestors joined the Freedom Convoy in Ottawa on January 29, 2022. There were reported instances in which the protestors harassed members of the local community, while a few carried Confederate flags or swastikas [46, 47]. Among the protest signs vilifying Canadian prime minister Justin Trudeau were a few that compared vaccine mandates to Nazi atrocities against the Jews. The truckers disrupted daily life in and around Ottawa, and the convoys blockaded international border crossings between Alberta and Montana in the West, as well as the Ambassador Bridge between Ontario and Michigan. These actions created large economic losses for companies that depend on cross-border or Canadian trucker supply chains—by some estimates reaching $500 million per day [46]. In the ensuing days, the number of protestors dwindled, and three weeks after it began, Canadian law enforcement broke up the final protests.

Several news outlets reported that far-right and conservative anti-vaccine activists committed to health freedom in the United States had important roles in either funding or encouraging the Freedom Convoy and Ottawa protests [48]. They included Rep. Marjorie Taylor Greene (R-Georgia) and retired army Lt. Gen. Michael Flynn, in addition to Fox News anchor Sean Hannity, who encouraged protest organizers on his broadcast by stating, "You do have a lot of support from your friends in America. That I can tell you" [48]. Hannity also indicated that similar protests would begin in the United States, while Sen. Rand Paul (R-Kentucky) expressed his hope that the truckers would "clog up cities" in America. On February 9, 2022, House Freedom Caucus Rep. Lauren Boebert (R-Colorado) tweeted: "Freedom is contagious and there's no vaccine that can shut it down. The Canadian Freedom Convoy has sparked a fire in the hearts of patriots. . . . Let's take our nation back from medical tyranny!" [49]. It is also relevant that some of the crowd-funding contributors were identified as prominent conservative or GOP donors based in the United States [48].

Two weeks after the Canadian protests ended, America's version of the Freedom Convoy, known as the "People's Convoy," comprising hundreds of trucks, motorcycles, and other vehicles, drove along the Capital Beltway in Washington, DC [50]. These protests continued to inconvenience DC-area commuters for several weeks, but law enforcement successfully blocked the People's Convoy from entering the nation's capital. However, similar efforts were also replicated in France and Belgium, where convoys of vehicles protesting COVID restrictions disrupted traffic in Paris and Brussels, respectively. At earlier times in the pandemic, more menacing health freedom protests and marches took place in London, Paris, Berlin, and other European capitals. For example, during the Berlin protests against COVID prevention measures in 2020, an attempt was made to storm the Reichstag, the German Parliament, with news outlets reporting ties that linked these activities to QAnon and several extremist groups [51]. In 2021 and into 2022, anti-vaccine and anti-mask rallies—and later, protests against "vaccine passports"—continued across western Europe [52]. It is interesting to note that many of those refusing vaccines in Germany voted for Alternative für Deutschland, a populist party linked to the far-right and opposed to both the European Union and immigration, while anti-vaccine activities fall along similar right-wing or populist lines in Italy, France, and Austria [53]. In France, populist and far-right activists led anti-vaccine protests in Paris and several other cities in 2021 and in 2022 fought vaccine passes or passports. These activities bore some resemblance to the 2018–19 "yellow vest" or "yellow jacket" protests in France against President Emmanuel Macron [54, 55]. In Austria, a new political party even formed around COVID-19 protests. Known as MFG (Menschen Freiheit Grundrechte or People Freedom Fundamental Rights), some in the local media refer to it as the vaccine-skeptics party [56]. In the summer of 2022, it was reported that anti-vaccine activists drove a prominent Austrian physician and staunch vaccine defender to suicide [57]. In the Netherlands, anti-COVID prevention and anti-vaccine protests took place in Amsterdam in January 2022, leading to multiple arrests [58]. Anti-vaccine activism is now connected with far-right extremist groups in Italy [59]. Through Telegram, Facebook, and other social media platforms that link to the Informare x Resistere (Inform and Resist) website, the Italian far-right claims vaccines and vaccinations are dangerous or a means by which to imprison people in their homes [60]. Elsewhere in southern Europe, vaccination rates in Albania have declined as a result of anti-vaccine rhetoric and activities [61], while anti-vaccine sentiments are rising in Greece [62].

Increasingly, such US-style anti-vaccine and anti-science activities based on health freedom propaganda may go beyond even Canada or Europe and now infiltrate low- and middle-income countries in Africa and elsewhere [63–67]. Vaccine hesitancy in these developing regions is a complex, evolving issue and the subject of an intense investigation led by Heidi Larson at the London School of Hygiene and Tropical Medicine [64]. The features of vaccine hesitancy and refusal in Africa and Asia, especially in impoverished rural areas, are often very different from the North American style of far-right health freedom protests that now also affect western Europe. However, it is interesting to note, for example, that in South Africa, the *Atlantic*'s Olga Khazan reported that vaccine hesitancy appears to be higher among whites than Blacks and that anti-vaccine material from the United States, including video clips of a Fox News anchor and Fauci memes, circulate there [53, 65]. She also notes that AfriForum, an NGO representing the interests of the Afrikaners, a segment of South Africa's white population, publicly opposes vaccine mandates [53]. In Uganda, openDemocracy, an international media platform, notes that American conservative groups have been operating to discredit Western vaccines, while the Center for Strategic and International Studies, a US policy think tank based in Washington, DC, finds that African communities have now been infiltrated by both QAnon and far-right talking points [65]. Furthermore, anti-vaccine activists based in the United States produced a video claiming that the tetanus toxoid used to immunize pregnant women is also inducing infertility among African women [66]. Previously, the chair of the Kenya Catholic Doctor's Association fought against introduction of the tetanus toxoid vaccine on this premise, as well as the human papilloma virus vaccine. When COVID-19 emerged, this same individual promoted the use of hydroxychloroquine before he too lost his life from the virus [67]. *Mother Jones* reports that Sen. Rand Paul's statements opposing COVID-19 vaccine mandates in the United States reportedly have infiltrated social media networks in Vietnam, while rhetoric against Bill Gates and the Gates Foundation, a strong supporter of vaccines, is pervasive in some African countries [68]. Therefore, we might expect health freedom propaganda to increasingly permeate the vaccine ecosystem on the African continent [65].

In other cases, it is uncertain whether anti-vaccine events reflect a global expansion of US-Canadian health freedom activism or rather a homegrown element. Prof. Heidi Larson, at the London School of Hygiene and Tropical Medicine, has surveyed vaccine acceptance across the globe—on every continent and almost every country. Her team reports on vaccine hesitancy in places as diverse as France, the Philippines, and now many African nations, showing that it comes in many different flavors and ideologies. Now, UNICEF, the Yale Institute of

Global Health, and the Public Good Projects have joined to establish a new Vaccination Demand Observatory, which includes a public dashboard of vaccine disinformation [68, 69].

This is an evolving situation, and my concern is that as the world becomes more connected, the US or North American far-right extremist viewpoint will become dominant or that it will merge with homegrown anti-vaccine activities. Since the start of the pandemic, multiple polio vaccinators have been assassinated in Afghanistan, including eight over a brief period in the first quarter of 2022, and nine the year before [70]. For those of us so committed to vaccines and vaccinations, few things are more heartbreaking. Many vaccinators in this part of the world are young and idealistic women, and all are absolute heroes. The story of vaccinator assassinations, some conducted by extremists in remote and isolated areas of Afghanistan and Pakistan, reflects political divisions as much as it does fundamentalist beliefs. On a Zoom event in March 2022 organized by the Consortium of Universities for Global Health, I was on a panel with Prof. Larson when she wondered what would happen when these women start giving COVID-19 vaccines. The anti-vaccine rhetoric arising from political extremism in the United States could eventually proliferate and bolster anti-science activities everywhere. The fact that this process is already under way on the African continent is an ominous sign. Anti-vaccine activism has globalized, and health freedom propaganda is now widespread in Europe and Africa. It is both potent and deadly.

Death rates are so high in the States—eye-wateringly high.

—Prof. Devi Sridhar, University of Edinburgh, *New York Times* interview

If I wanted to guess if somebody was vaccinated or not and I could only know one thing about them, I would probably ask what their party affiliation is.

—Liz Hamel, Kaiser Family Foundation, National Public Radio interview

3 | Red COVID

In 2021, the third year of the pandemic, the deaths from COVID-19 really began to climb. Initially it was a terrible wave from the Alpha variant in the winter, followed by a summer–fall Delta wave. The Delta variant of the SARS-2 coronavirus was especially rough owing to the high toll it took in terms of deaths and illness. It predominantly affected the southern US states and Texas. By Christmas, more than 800,000 Americans had lost their lives to the virus. At times my sadness during the last half of 2021 became overwhelming, and I remember tearing up on several occasions while I was being interviewed on CNN and MSNBC. For a while it became difficult not to become emotional when delivering such profoundly sad news. At first, I worried how television audiences might respond, but eventually two things began to happen. First, maybe because the grief was truly overwhelming, I stopped caring as much about how my emotions were understood, and second, the responses I received on social media and from family and colleagues were mostly positive. It turned out that the American people appreciated the fact that I was a physician-scientist who actually cared enough to express grief and related emotions.

Later, when I realized that so many deaths occurred among those who had refused to be vaccinated, I experienced an additional emotion—anger. I grew angry at those pushing health freedom propaganda and disinformation about the pandemic and vaccinations. It became obvious to me that some elected leaders at both the state and federal levels had a hand in encouraging unnecessary deaths. Moreover, I felt that many politicians who endorsed an anti-vaccine agenda did so not out of ignorance but for reasons of partisan expediency. When I began expressing my disgust and anger toward those willing to sacrifice American lives for political gain, that too caused many viewers (judging by the e-mails and notes on social media I received) and journalists (judging by the interview requests following a cable news appearance) to take notice. I will never forget the year 2021.

The United States: "Ground Zero" for COVID-19 Deaths

For almost the entire pandemic, the United States has been ground zero for COVID deaths. Not only has its death toll from the virus now exceeded one million, but the country has also suffered more COVID deaths per capita than any other large-population, high-income nation [1]. According to a global analysis of COVID-19 deaths in 2020–21 by the University of Washington's Institute for Health Metrics and Evaluation (UW-IHME), the United States was second only to India in numbers of deaths [2].

Unlike other high-income nations, the trajectory of COVID-19 deaths in the United States is unique. Whereas a majority of the deaths in western European nations and Canada occurred before the arrival of COVID vaccines, two of the leading tracking centers (UW-IHME and Johns Hopkins University), as well as the *New York Times* report that deaths in America continued to climb precipitously even after vaccines were made widely available to the public [3, 4].

In the spring of 2021, the Biden White House announced that all Americans would become eligible to receive a COVID-19 vaccination by May 1, 2021 [5]. In the United Kingdom, a similar target was set for July 1, 2021 [6]. If we benchmark those time points, we see a stark contrast between the United States and other advanced nations. For example, the UW-IHME finds that only 20% of deaths in the United Kingdom occurred after vaccines were widely distributed in 2021 and up until the start of spring the following year [3]. Or looking at it another way, only 13% of the UK deaths occurred between July 1, 2021, and the end of that year. In contrast, in the United States approximately 40% of the deaths occurred after vaccines were fully distributed. This includes 245,000 Americans who lost their lives between May 1 and December 31, 2021, according to the UW-IHME. Across multiple regions of the United States, the deaths continued to accelerate throughout the last half of 2021 and into the following year. In my state of Texas, for example, almost one-half of its 90,000 deaths happened after May 1, 2021 [7]. These deaths occurred despite the fact that almost any adult and teenager was eligible to receive an mRNA vaccination, and vaccines were freely available.

Thus, whereas COVID-19 deaths and death rates in the United Kingdom and advanced western European nations began to level off after the spring–summer of 2021, in the United States things were just heating up. The deaths (on a daily basis) resulting from the pandemic, based on data from the Johns Hopkins University Coronavirus Resource Center, are shown in figure 3.1.

The deaths in America in 2021–22 came in three major tranches. First, in the winter of 2021, the Alpha wave took a heavy death toll. At the time, vaccines were just beginning to be released to the American population, beginning with seniors. The arrow in the figure points to May 1, 2021, when COVID-19 vaccines became fully and freely available, just ahead of the following Delta variant wave. Beginning in the

THE DEADLY RISE OF ANTI-SCIENCE

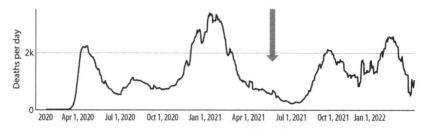

Figure 3.1. Daily deaths from COVID-19 in the United States, early 2020 to mid-2022. The arrow points to May 1, 2021, the date when COVID-19 vaccines were made widely available. The peaks of daily deaths to the left of the arrow represent those that resulted from the original lineage in the spring and summer of 2020, followed by the Alpha variant that was introduced from the United Kingdom or western Europe in the winter of 2021. The two major peaks to the right of the arrow represent the Delta variant in the summer and fall of 2021, mostly across Texas and the southern states, followed by the Omicron variant that began in late fall and continued into the winter of early 2022. *Source*: Modified from Johns Hopkins University School of Medicine, Coronavirus Resource Center, https://coronavirus.jhu.edu/region/united-states.

summer and lasting until the fall, the Delta variant caused thousands of deaths among unvaccinated populations in Texas and other southern states. Reaching 2,000 deaths per day, the mortality rate in America during the Delta wave was vastly higher than that in the United Kingdom. Texas had perhaps the greatest number of Delta deaths (around 25,000) [3, 7], with 85% among the unvaccinated, according to the Texas Department of State Health Services [8].

Then, after the Delta wave subsided, an equal number of deaths resulting from the Omicron variant (the third major tranche) occurred in the winter months of 2022. By the time Omicron arrived in America, just a few weeks after it emerged in the United Kingdom, it caused the "eye-wateringly high" numbers of American deaths noted in the epigraph to this chapter—exceeding 2,500 per day [1]. In fact, the Omicron wave death rate in America was almost as high as the previous worst wave, which had begun toward the end of 2020 when the Alpha variant entered the country before vaccines were widely available. In contrast, in the United Kingdom deaths remained in the 200–300 per day range.

44

The unique and aggressive rise in COVID-19 deaths in America mostly reflected an inability to achieve high vaccination coverage despite an abundance of vaccines. The CDC found that, overwhelmingly, the COVID deaths during the last half of 2021 occurred among the unvaccinated, as shown in figure 3.2. During the period of Delta predominance (July to November), the incidence rate ratio comparing unvaccinated patient deaths to fully vaccinated patient deaths was approximately 16.3 to 1 [9]. However, this ratio was as high as 21.9 to 1 prior to Delta, when the Alpha variant predominated.

Although the United States was one of the first nations to receive access to mRNA vaccines from Pfizer-BioNTech and Moderna (toward the end of 2020), eventually amassing an impressive arsenal of these technologies, vaccination rates were consistently high only in the initial vaccine rollout during the first few months of 2021. By April and

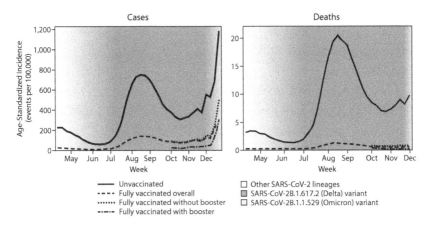

Figure 3.2. Weekly trends in age-standardized incidence of COVID-19 cases (April 4–December 25, 2021) and deaths (April 4–December 4, 2021) for unvaccinated compared with fully vaccinated persons, overall and by receipt of booster doses and national weighted estimates of variant proportions—25 US jurisdictions. *Source*: Johnson AG, Amin AB, Ali AR, et al. COVID-19 incidence and death rates among unvaccinated and fully vaccinated adults with and without booster doses during periods of Delta and Omicron variant emergence—25 U.S. jurisdictions, April 4–December 25, 2021. MMWR Morb Mortal Wkly Rep 71 (2022): 132–38. https://www.cdc.gov/mmwr/volumes/71/wr/mm7104e2.htm.

May and for the remainder of the year (and into 2022) vaccination rates stalled. Then, when the need for booster shots became apparent in the fall, the US government and its major federal agencies, including the CDC, failed to persuade most Americans to get one. By the time the Omicron wave hit, only one-quarter of the nation was fully vaccinated and boosted, far below the rates of the western European nations and Canada. At the peak of the Omicron wave, the COVID-19 death rate in the United States was several times those of Germany or the United Kingdom [1].

To be sure, the high rates of obesity, diabetes, hypertension, and other comorbidities linked to poor COVID-19 outcomes among the American population further contributed to the higher death rate. This is especially true in many southern states and Texas, where COVID-19 death rates are particularly high [7, 10]. However, the fact that the trajectory of COVID-19 deaths more or less overlaps with European nations up until the time vaccines were widely available suggests that stalling vaccination rates accounted for most of the deaths. Somehow, low vaccination coverage and sky-high COVID-19 death rates became normalized in America.

Ultimately, our national failure to vaccinate Americans was responsible for a high proportion of the 245,000 COVID-19 deaths from May 1 to December 31, 2021 (as estimated by UW-IHME) [3]. Most of these deaths occurred during the wave resulting from the Delta variant that began in June, accelerated in the summer and fall, and wound down toward the end of the year. Based on a Peterson Center on Healthcare and Kaiser Family Foundation (Peterson-KFF) analysis of the percentage of COVID deaths constituted by unvaccinated Americans at the peak of the Delta wave in August 2021 (81%) and September 2021 (79%), this lack of vaccination could account for approximately 196,000 American lives lost [11]. Another estimate from Peterson-KFF places that number closer to 162,000 for the period between June 1 and the end of 2021, but 234,000 if the first three months of 2022 are included [11]. Still another way to estimate the deaths is to use the CDC's published 16.3 to 1 ratio of unvaccinated to vaccinated deaths during the July to November 2021 Delta wave to derive approximately 230,000 deaths [9]. A slightly lower number of 208,000 deaths can be obtained using the Texas Department of State Health Services figure (85% were unvaccinated) [8]. Therefore, of the 245,000 deaths, roughly 200,000 occurred among those who chose not to take a COVID-19 vaccine.

Summary and Derivation of Approximately 200,000 American Lives Lost

The estimate of 200,000 deaths is calculated by counting the 245,000 cumulative COVID-19 deaths in the United States between May 1 and December 31, 2021, according to the University of Washington–Institute for Health Metrics and Evaluation's "COVID-19 Projections" [3], multiplied by a factor of 0.80. The 0.80 number reflects the percentage of deaths constituted by unvaccinated Americans at the peak of the Delta wave in August 2021 (81%) and September 2021 (79%) according to a Peterson-KFF analysis of CDC information [11] to account for 196,000 lives lost. The CDC estimates a ratio of 16.3 to 1 of unvaccinated to fully vaccinated deaths during the period of Delta predominance (July to November 2021 [9, table 1]), equivalent to 230,000 deaths. Another approximation reflects the percentage of deaths made up by the unvaccinated (85%) versus those partially or completely vaccinated (15%) for the state of Texas for the year 2021 according to the Department of Texas State Health Services [8], which works out to 208,000 lives lost. The health analyst Charles Gaba estimates between 180,000 and 235,000 deaths [12]. Peterson-KFF also provides an additional (and somewhat lower) estimate of 162,000. This lower estimate is based on the 91% effectiveness versus death caused by the Delta variant for the two mRNA COVID-19 vaccines [11]; therefore, even if all of the unvaccinated had accepted their vaccines, there would still be some deaths. Peterson-KFF also find that vaccines would have prevented 234,000 deaths if we include the first three months of 2022 [11].

Other independent estimates generally support my analysis. The health analyst Charles Gaba, who created ACASignups.net, determined that 180,000–235,000 Americans have died since May 1, 2021, because they refused COVID-19 vaccinations [12]. An epidemiology research group at Harvard's T.H. Chan School of Public Health estimated that 135,000 excess deaths occurred among unvaccinated Americans in a period between May 30 and December 4, 2021, among adults over the age of 18 [13]. However, this figure does not include a significant number of deaths also occurring in the last month of the year.

Demographics of the Unvaccinated and the 200,000 American Deaths

What do we know about those who refused vaccinations—those who needlessly lost their lives even though their deaths could have been prevented had they accepted COVID-19 vaccines? The answer is complicated by an apparent shift among those who refused vaccinations right after mRNA vaccines were first rolled out to the general public near the end of 2020, versus the unvaccinated in the last half of the following year, when anyone who wanted to get vaccinated could do so. Early in the pandemic several groups stood out either for their vaccine hesitancy or because they lacked access to the vaccines. By May 3, 2021, the KFF found that Asians represented the group with the highest COVID-19 vaccine coverage (with at least one vaccine dose) at 48%, followed by Whites (39%), Hispanics (27%), and Blacks (25%) [14]. In the ensuing months, however, those racial and ethnic vaccine equity gaps gradually closed, such that by April 2022, Whites, Blacks, and Hispanics had accepted COVID-19 vaccinations at similar levels, roughly between 57 and 65%, although these rates still lagged significantly below Asians at 85%. The reasons why racial and ethnic vaccine equity gaps in America narrowed are still under investigation through several efforts, including the Rockefeller Foundation's Equity-First Vaccination Initiative, formerly under the direction of Otis Rolley [15].

My own personal experiences are relevant. When the early pandemic polls began showing significant rates of vaccine hesitancy and refusal among Hispanic and African American groups, I committed to going on regular podcasts, public appearances, and radio interviews to reach those populations. I testified before the Congressional Hispanic Caucus and also did regular interviews on Spanish-language television stations in the United States such as Univision and Telemundo. For a while, I did almost weekly virtual interviews with Deysy Rios, one of the Houston-based Univision reporters. My only regret about those interviews is that I did not feel very confident in my Spanish language skills, so we conducted them in English. During the pandemic I also participated in regular Zoom meetings with African American groups and appeared on multiple radio shows and podcasts that reached Black audiences. Among them were HoodMed Chats, Roland Martin Unfiltered, and Texas Southern University KTSU radio. Through Hood Medicine, headed by Dr. Shanice Hudson and her colleagues Jonathan White, Doug Slaughter, and others, I got to meet some extraordinary science and healthcare professionals committed to vaccine equity in Black communities [16]. In time, they became friends and supportive colleagues.

One Zoom interview that stands out in my mind was arranged by Dr. Eric Freeman, an African American pediatrician in Richmond, Virginia, who organized an event with members of the Providence Park Baptist Church. The church pastor explained that reductions in vaccine hesitancy depended on doctors who were helpful in speaking out. It made a deep impression when he told me how an informal network of pastors at African American churches worked hard to convince people about the effectiveness and safety of mRNA vaccines. The minute he said this, something clicked, and I thought to myself "this makes total sense." I believe the Black clergy stepped up in a big way to promote vaccine equity in the United States.

However, as vaccine hesitancy decreased among people of color, there was no progress in convincing another group to vaccinate—labeled in various surveys as "conservatives" or "Republicans." In time, they became the dominant group resisting and refusing COVID-19 vaccinations in America. In 2020, I collaborated with a group of social scientists based at Texas A&M University School of Public Health, led by Dr. Tim Callaghan (now at Boston University). The studies confirmed high rates of vaccine hesitancy among the African American community; however, the most striking finding was the strong association between vaccine hesitancy or refusal with those self-identifying as conservatives or Trump voters [17]. For me, this was not unexpected, given my firsthand experiences with conservative groups linked to health freedom, but now we had actual social science data to support this dangerous and emerging trend during the COVID-19 pandemic.

A subsequent KFF poll conducted in October 2021 found that slightly more than one-quarter of adults in America refused to get a COVID-19 vaccine, but 60% of those refusing self-identified as Republican or Republican-leaning, compared to 17% of Democrats [18]. Kaiser also found that those unvaccinated Republicans tended to be more conservative, younger (18–50 years old versus those over 50), and less educated (high school or less) than vaccinated Republicans. Most tellingly, 96% of the unvaccinated Republicans indicated that vaccination is a personal choice, touting a central tenet of health freedom. In addition, almost as many who said they would not accept a vaccine dismissed the severity of COVID-19 as "exaggerated." This contrasted with vaccinated Democrats, who overwhelmingly felt that COVID-19 was a very serious illness and that getting vaccinated in order to protect others was an important social responsibility. Still another aspect worth noting was the higher percentage of the unvaccinated living in rural versus urban areas of the country. A second poll conducted in August 2021 by NBC News confirmed the Kaiser findings. It concluded that the groups with the lowest vacination rates were Republicans (only 55% vaccinated), those living in rural areas (52% vaccinated), Trump voters in the

2020 election (50% vaccinated), and "Republicans who support Trump more than party" (46% vaccinated) [19].

Finally, a Pew Research Center survey conducted in early 2022 found similar stark differences between Republicans and Democrats in terms of those who were willing to get immunized against COVID-19 or receive a booster [20]. The least-vaccinated group consisted of less-educated Republican adults, meaning those with a high school diploma or less. A study led by University of Colorado social scientists also noted how lower educational attainment and health literacy predicted susceptibility to health misinformation, although this particular analysis was in the context of the human papillomavirus vaccine and other health interventions [21]. Therefore, a picture emerged in which younger, less educated, conservative Republicans accepted health freedom propaganda (which first arose out of Texas in the 2010s) and incorporated it into their health decisions. In early 2022, the UW-IHME attempted to map the geographic distribution of the greatest vaccine hesitancy or refusal (fig. 3.3).

Not surprisingly, high levels of vaccine hesitancy occur in areas where one might expect a high concentration of individuals who meet the younger, more conservative, less educated, and Republican profiles identified by KFF, NBC News, and the Pew Research Center. Such regions might include the southern US and Texas, especially west and central Texas, including the panhandle region and adjacent areas of Oklahoma; or east Texas and adjacent areas of Arkansas, Louisiana, and Oklahoma. The Gulf Coast areas of Texas, Louisiana, Mississippi, and northern Florida also stand out. Even before the pandemic, I wrote, together with Rep. Sheila Jackson Lee (D-Texas), about how the Gulf Coast was vulnerable to infectious diseases owing to a confluence of social and physical determinants, including extreme poverty, human migrations, and climate change [22]. Many of these areas bore the brunt of the Delta wave of COVID-19 in terms of hospitalizations and deaths. The UW-IHME further reported that vaccine hesitancy is high in southern Missouri, the Appalachian region, and in parts of the Mountain

THE DEADLY RISE OF ANTI-SCIENCE

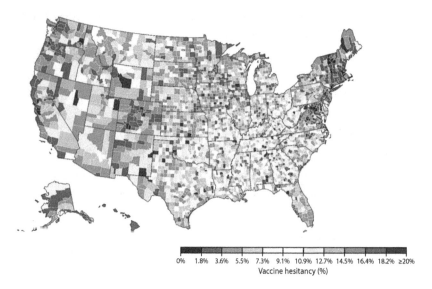

Figure 3.3. Percentage of respondents who indicate some vaccine hesitancy, by county, February 18–24, 2022. The original, colorized image is best viewed online, where the color key is available, https://www.healthdata.org/data-visualization/covid-19-vaccine-hesitancy and also at https://www.press.jhu.edu/books/title/33293/deadly-rise-anti-science. *Source*: Institute for Health Metrics Evaluation. Used with permission. All rights reserved.

West. Each of these areas trend toward conservatism or are considered Republican-majority states and counties.

New National Trends

An overarching concept of "red COVID" (red referring to the color arbitrarily assigned to the Republican Party by the *New York Times* in 2000, versus blue for Democrats) is based on a convergence of studies from KFF, Pew Research Center, and Charles Gaba, together with the *New York Times*, *Axios*, and National Public Radio (NPR). On October 1, 2021, David Leonhardt published a revelatory *New York Times* article titled "Red Covid" for his morning newsletter series [23]. While many of us who followed vaccine hesitancy closely already knew about the red/blue split in vaccine acceptance, Leonhardt, through a set of compelling graphics, informed the public about a striking partisan and geographic divide over vaccination as we entered the last half of 2021.

In stark terms, the idea of "red COVID" points out that as the United States entered the last half of 2021, each blue, liberal state with a Democratic majority that had voted for Joe Biden in the 2020 election had achieved far higher vaccination rates than conservative, red states that had voted for Donald Trump. Leonhardt states plainly: "The political divide over vaccinations is so large that almost every reliably blue state now has a higher vaccination rate than almost every reliably red state" [23]. Given the high rate of protection that vaccines confer against hospitalization and deaths, especially versus the original lineage of the virus and its Alpha and Delta variants, those low vaccination rates correlated with high death rates. Therefore, it was no surprise (although still upsetting) that COVID-19 deaths also demonstrated a similar pattern of partisan split. Importantly, the differences in vaccination rates represent much more than an abstract concept; they directly translate into a partisan division in terms of lives lost. By the summer and fall of 2021, as the highly transmissible Delta variant became dominant, overwhelmingly those losing their lives from COVID-19 were living in conservative counties with a majority that voted for Trump the in 2020 (fig. 3.4).

In a follow-up analysis a month later in November, the *New York Times* reported that "heavily Trump" counties exhibited a death rate from COVID-19 three times higher than "heavily Biden" counties [24]. These red COVID losses of life aligned with the polling from KFF and NBC News, as well as Dr. Callaghan's paper that highlighted increased death rates among Republicans, especially those who were younger and less educated. For example, according to the *New York Times*, Charles Gaba found that in counties where 70% or more voted for Donald Trump, almost 50 per 100,000 people died from COVID-19, whereas only 10 per 100,000 died in the counties where only 32% or less sought to reelect President Trump [23]. Moreover, the higher the percentage of

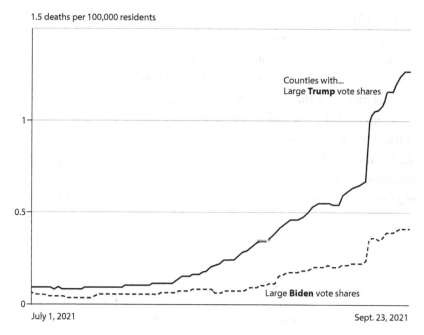

Figure 3.4. Average daily COVID-19 deaths in the United States, based on 14-day averages, July–September 2021. *Source*: Leonhardt D. Red Covid. New York Times, September 27 (updated October 1), 2021, https://www.nytimes.com/2021/09/27/briefing/covid-red-states-vaccinations.html. © 2021 The New York Times Company. All rights reserved. Used under license.

Trump voters (meaning the redder the county), the greater the loss in life [23, 24]. In another analysis from October 2021, Charles Gaba found that the death rate was more than sixfold in the upper one-tenth of Trump-voting counties compared with the lowest one-tenth [25]. A later analysis from February 2022 finds that a "high-end estimate would mean roughly 70% more Trump voters than Biden voters have died of COVID since the 2020 election," according to Gaba [26]. The association is so dramatic that Liz Hamel, vice president of public opinion and survey research at Kaiser, stated, "If I wanted to guess if somebody was vaccinated or not and I could only know one thing about them, I would probably ask what their party affiliation is" [25].

Another representation of the partisan divide in vaccination rates, which Charles Gaba lines up with Democratic/Republican leanings, is depicted in figure 3.5. The findings confirm a sharp blue versus red state divide among the more than 3,000 counties in the United States [27]. In

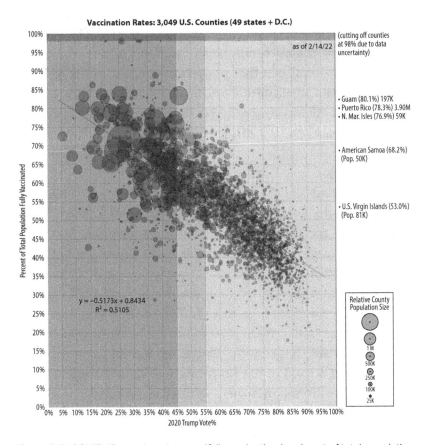

Figure 3.5. COVID-19 two-dose (or one J&J) vaccination levels out of total population by county, as of February 14, 2022. Dark shading to the left indicates blue-Democratic, and light shading to the right indicates red-Republican. N. Mar. Isles = Northern Mariana Islands. The original (colorized) image can be best viewed at https://acasignups.net/22/02/14/weekly-update-county-level-covid19-vaccination-levels-2020-partisan-lean and also at https://www.press.jhu.edu/books/authors/peter-j-hotez-md-phd. *Source*: Gaba C. Weekly update: County-level #COVID19 vaccination levels by 2020 partisan lean. ACASignups.net, February 14, 2022.

many counties where the vote for Trump exceeded 75%, less than one-half and in many cases only 35–40% of the population is vaccinated. This has been my experience as well with conservative counties in central and east Texas, along with the panhandle. The Gaba graph exhibits additional interesting features, including the large circles for counties represented in the highly vaccinated, low-Trump-voting areas versus generally smaller circles in the high-Trump-voting areas. This suggests that urban counties are generally better vaccinated than rural counties, an urban versus rural split that was predicted based on the KFF and other survey data highlighted above. In addition, the *New York Times* notes that some conservative writers or thought leaders have attempted to explain the sharp partisan divide on the basis of confounding factors, such as age or weather [24], but such associations have been dismissed by most experts.

Another analysis of the partisan divide in COVID-19 deaths conducted by Charles Gaba is shown in figure 3.6. Here, he looks at the multiple peaks and valleys of deaths during the pandemic from 2020 until the first quarter of 2022. The initial wave that hit New York City and the Northeast in April–May 2020 was of course in Democratic strongholds, but from then onward, Republican-dominated counties suffered disproportionately from COVID-19 deaths. Most dramatic were the summer–fall waves in 2020 and 2021 from the original lineage and Delta variant, respectively. Except for the initial peak, when COVID-19 first arrived in the Northeast, COVID-19 hospitalizations and deaths have disproportionately occurred in red states.

In a related metric, there were many COVID-19 deaths that were assigned inappropriately to other causes because the deaths occurred at home or were not properly certified by coroners and medical examiners who were overwhelmed at the worst moments of the pandemic and took the family's word about the actual cause of death [28]. Measuring "excess deaths" above what might be considered a typical level offers an approach for identifying an additional 170,000 COVID-19 deaths beyond the official toll, with findings that these ran 21 times higher in the

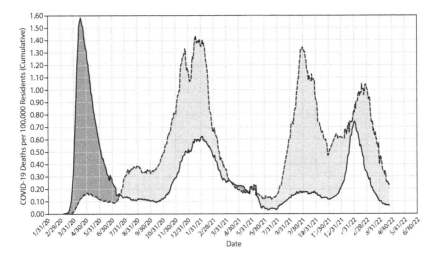

Figure 3.6. US COVID-19 deaths / 100,000: County-level reddest 10% vs. bluest 10% (14-day moving average since the start of the pandemic). The dark shading between the two lines in the first peak indicates that the upper line is blue, showing that the death rate was much higher in strongly Democratic areas; and the light shading in the other three peaks indicates that the upper line is red, showing that the death rate was much higher in strongly Republican areas. The original (colorized) image can be best viewed at https://acasignups.net/22/04/25/monthly-update-covid-death-rates-partisan-lean-vaccination-rate and also at https://www.press.jhu.edu/books/authors/peter-j-hotez-md-phd. *Source:* Data from Johns Hopkins University, New York Times, White House COVID Response Team. Graph from Charles Gaba @charles_gaba / ACASignups.net.

reddest counties, as measured by percentage of Trump voters, compared with the bluest counties [28]. These data from Charles Gaba in collaboration with a group of public health experts at Boston University and other institutions, led by Dr. Andrew Stokes, are presented in figure 3.7 and again emphasize the stark divide in both COVID and non-COVID excess deaths.

The irony of losing so many Republican voters needlessly to COVID-19 has not been lost on some leaders of the GOP. So many unvaccinated Republican voters died from COVID-19 that John Nolte of the Breitbart far-right news outlet pointed out, "Right now, a countless number of

THE DEADLY RISE OF ANTI-SCIENCE

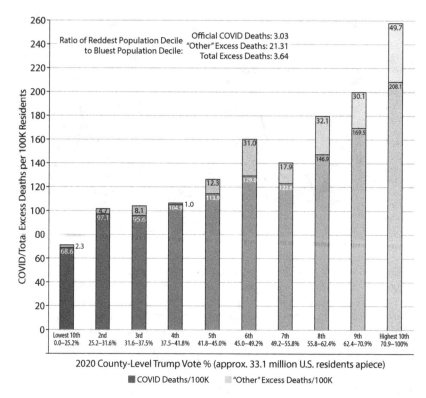

Figure 3.7. COVID versus non-COVID excess death rates, January–December 2021, by Trump 2020 county-level vote %. Shading to the right corresponds to red (Republican) in the original image, versus blue (Democratic) to the left. The original (colorized) image can be best viewed at https://acasignups.net/22/05/09/exclusive-non-covid-excess-death-rates-ran-21x-higher-reddest-counties-bluest-2021 and also at https://www.press.jhu.edu/books/authors/peter-j-hotez-md-phd. *Source*: Gaba C, Stokes A. "Non-COVID" excess death rates ran 21x higher in reddest counties than bluest in 2021. ACA Signups, May 5, 2022.

Trump supporters believe they are *owning* the left by refusing to take a lifesaving vaccine.... In a country where elections are decided on razor-thin margins, does it not benefit one side if their opponents simply drop dead?" [quoted in 23]. He has a point. Merely a few tens of thousands of Democratic votes in the states of Arizona, Georgia, Nevada, and Wisconsin *combined* prevented the reelection of former president Donald

Trump, and the loss of so many Republican voters could factor into future election outcomes [29], including possibly the recent November 2022 midterm elections. In his article "How Republicans Failed the Unvaccinated," *New York Times* columnist Ross Douthat laments how the GOP leadership squandered an opportunity to save thousands of Americans lives [30]. On more than one occasion, I have asked myself, how can this make sense at any level?

Shifting Trends in the Omicron Wave

As the Omicron variant began to accelerate during the first quarter of 2022, several observers noticed that while COVID-19 continued to kill far more people in red states than in blue ones, this gap began to decrease somewhat from what had been documented during the previous Delta wave [31, 32]. Possibly this shift reflected the partial immune-evasion properties of Omicron so that even fully vaccinated (but not boosted) individuals were succumbing to COVID-19. This was especially true for those who were immunocompromised, extremely old, or had several comorbid conditions. For example, almost two-thirds of those who lost their lives during the Omicron wave were 75 and older, including many who were vaccinated. However, many of those vaccinated deaths were in individuals who did not get boosted despite their eligibility for a third immunization. The *Washington Post* reported that in two states it analyzed, California and Mississippi, three-quarters of the vaccinated seniors who died from COVID-19 had also failed to receive boosters [33]. The new reality during the Omicron wave was that getting boosted became an essential component of fending off severe illness from COVID-19, in addition to efforts to reduce virus exposures through masks and social distancing. For reasons that I never understood during this period, the CDC still continued to define two mRNA vaccine doses as constituting "fully vaccinated" criteria. For me, it had become increasingly clear that a boost was needed to best fend off hospitalizations and deaths, especially for those over the age of 50. This was even confirmed by data from the CDC [34].

Another reason for the closing gap is the possibility that some unvaccinated individuals who had survived previous bouts of COVID-19 may have received some partial immunity that prevented them going into the hospital or dying from Omicron. Still another factor may have been the increasing availability of effective antiviral drugs such as Paxlovid from Pfizer or newly available monoclonal antibodies. Thus, some individuals who bypassed vaccinations may have had access to COVID-19 medications that reduced their likelihood of hospitalization or death.

Finally, even in the blue regions there remain many individuals who continue to refuse vaccinations. Conservative voices and their representatives in legislatures are found in every state. In solidly blue New Hampshire in 2022, its House of Representatives approved by a slim margin legislation to block federal vaccine mandates in its state-run medical facilities [35]. As noted above, the vaccine defiance against Washington governor Jay Inslee occurred among state troopers in a mostly blue state. However, their numbers are small compared with solidly large, red states such as Florida or Texas. Therefore, it comes as no surprise that an October 2022 study conducted by the National Bureau of Economic Research finds that the excess death rate for Republicans was higher than the excess death rate for Democrats. The NBER report concludes: "The gap in excess death rates between Republicans and Democrats is concentrated in counties with low vaccination rates and only materializes after vaccines became widely available" [36, 37].

A Special Responsibility

How can we address red COVID and halt the senseless loss of human life among Republicans and conservative-leaning individuals? That 200,000 unvaccinated Americans gave up their lives needlessly through shunning COVID-19 vaccines can and should haunt our nation for a long time to come. Having lived and worked in a red state such as Texas for the past decade, my sense of loss is especially acute. The state has suffered enormous losses because of the virus. By some estimates, 90,000 Texans have lost their lives to this illness [7, 38], including more than 20,000 during the last half of 2021 and up to 40,000 overall since May 1, 2021 [7]. Such numbers identify COVID-19 as one of our state's absolutely worst human tragedies. For comparison, an estimated 22,000 Texans sacrificed their lives for our country in World War II [39], while our worst natural disaster, the Galveston hurricane of 1900, took the lives of an estimated 6,000 residents [40].

I sometimes ask myself whether I could have done more to push the state leadership to promote vaccinations or to advocate for them myself. It was truly heartbreaking to see so many lives lost in my state and so much suffering that could have been prevented. I remind myself that I did several local television news interviews for Houston, Dallas, and Austin, as well as Spanish-language stations, practically every week of the pandemic (and usually several times weekly) to convince Texans about the safety and effectiveness of vaccines—not to mention regular interviews for all the local papers and magazines. On a frequent and regular basis, I would do full-page interviews with the *Houston Chronicle*, initially with Lisa Gray (now with City Cast Houston) and later with Andrew Dansby. I also did extensive interviews with the *Texas Tribune* and *Texas Monthly*, and countless Zoom or in-person interviews with school officials in the state, as well as employers, churches, town halls, and service clubs. One of my consistent messages was that everyone is entitled to their conservative views, and even extreme conservative views, but please do not adopt vaccine refusal or defiance. Health freedom propaganda is just that—fabrications—and you do not need to embrace this as part of your worldview. I stressed the urgency of decoupling COVID-19 prevention, especially vaccines, from the GOP or far-right.

Along those lines, in 2021, I stressed to the Biden White House COVID-19 Response Team that they could do more to reach out to the middle of the country. In my frustration, I would sometimes explain to them that their efforts resonate well enough along the Acela train corridor between Washington and Boston, and on the West Coast, but not necessarily in the central states. I recommended that President Biden establish some type of ambassador or outreach program composed of physician-scientists and epidemiologists who live and work in red states. I offered to assist in this role for my state of Texas. However, this approach has so far met with little or only very modest enthusiasm. Still another approach is to bring in members of the clergy, based on past successes by Black churches, as discussed above. A 2022 study conducted by the Values and Beliefs of the American Public Survey for the Gallup Organization and Baylor University (I was not involved in the study) found a strong association between "conservative religious beliefs" and the GOP [41]. Still others have pointed out that the term "conservative" does not adequately describe those defying vaccinations in red states. An older term, "Middle American Radicals," was coined in the 1970s to describe a group living far from the coasts who had no college education and an income in the middle- or lower-middle range. They are bound by a common grievance of feeling under siege and without a voice in government [42]. In some respect, these may prove to be apt descriptions of those at risk for refusing vaccines and susceptibility to COVID-19.

We might also consider the benefits of bringing in expertise from outside the health sector. I now cochair with Saad Omer from the Yale Institute of Global Health a Lancet Commission on Vaccine Refusal, Acceptance, and Demand in the USA to look at best practices and approaches on how to diversify the messengers for vaccinations and to expand research opportunities across the areas of social, behavioral, and communication science to shape a comprehensive strategy to respond to this and future pandemics [43, 44]. In parallel, I have suggested to the Biden administration the creation of an interdisciplinary task force of experts from departments such as Homeland Security, Commerce, Justice, and others in recognition of the fact that the loss of human life on this scale, as a result of partisan politics and defiance, is far bigger than what can be managed only by the Department of Health and Human Services. To date, there are no efforts planned to hold congressional hearings on the origins of vaccine refusal leading to this American tragedy. Certainly, there is no enthusiasm for creating an entity that resembles a truth-and-reconciliation commission at the national level similar to efforts made in post-Apartheid South Africa during the 1990s, in order to identify those individuals or groups who encouraged vaccine defiance.

At more than 200,000 dead Americans and counting, the stakes are very high. Based on these numbers alone, we must consider how vaccine refusal or defiance became a leading cause of losses in human life, especially among young and middle aged adults. Such numbers exceed the annual deaths from conditions such as breast and pancreatic cancer in the United States. More Americans died from COVID-19 vaccine refusal than diabetes or Alzheimer's disease, based on data for those conditions according to the CDC [45]. An estimated 200,000 Americans also die annually from accidents and unintentional injuries. Therefore, we might consider reframing the deaths in America attributable to vaccine refusal in the context of other conditions and recognize this as a cause of injury or illness, which is potentially preventable. To begin understanding the causes of vaccine refusal or defiance, it becomes important to look at those individuals or groups generating the political content that caused so many to needlessly lose their lives.

Now, they're starting to talk about going door to door to be able to take vaccines to the people.... Think about the mechanisms they would have to build to be able to actually execute that massive of a thing, and then think about what those mechanisms could be used for. They could then go door to door and take your guns. They could go door to door and take your Bibles.

—Rep. Madison Cawthorn (R-North Carolina), CPAC Summit, Dallas, Texas, July 2021

I think Fox has been almost single-handedly responsible for the politicisation of public health in the US and the creation of vaccine hesitancy in a significant portion of the population.... It's been tremendously damaging.

—Joseph Azam, former senior vice president at News Corp in New York

4 | An Anti-science Political Ecosystem

As the COVID-19 pandemic accelerated from 2020 to 2022, the health freedom propaganda that rose out of Texas as a protest to school-mandated pediatric vaccinations gained strength in terms of the number of its adherents, its organization, and a new, expanded scope. In time, health freedom became a national rallying cry against masks, social distancing, and ultimately COVID-19 vaccinations. Thousands of Republicans, especially those with lower educational attainment, and many living in rural areas of the southern United States, began demonstrating their allegiance to these tenets by refusing immunizations with any of the three US-approved COVID-19 vaccines.

Together with like-minded colleagues committed to vaccination, I viewed the events building over the summer and fall of 2021 with alarm, as deaths among the unvaccinated mounted rapidly during the Delta wave that was sweeping across the southern United States. I had a front-row seat to this entire horror show, living and working in southeastern

An Anti-science Political Ecosystem

Texas and the Gulf Coast—ground zero for where the deaths were among the highest. Intensive care units were overwhelmed as healthcare professionals appeared in tears on our local news channels, as did the stories of patients who expressed extreme remorse for refusing a COVID-19 vaccination. Many also died, however, insisting with their final words that COVID-19 was a hoax or some form of liberal conspiracy. I could not decide which one of these expressions was more upsetting.

As someone who closely followed the evolution of the anti-vaccine movement from its autism assertions to the defiance of immunization along a partisan divide, I became deeply curious about the external drivers that caused so many Americans to lose their lives. Many of the factors driving down vaccination in the red states became evident to me from my unique position as a commentator on the cable news channels. On multiple occasions the network anchor would ask me to reflect on the recorded public statements of an elected official, frequently a member of the US Senate or House of Representatives, or one of the governors or those serving in state legislatures. The political leaders representing the red states hit the hardest by COVID-19 were often the most misinformed. I was struck by how many of these elected officials made similar or identical points that dismissed either the effectiveness or safety of COVID-19 vaccinations. I began to understand that something deliberate and well-organized might be under way. Could their anti-vaccine talking points represent coordinated disinformation from the far-right? Moreover, I would routinely see their public statements amplified by Fox News and other conservative news outlets in the evening or on the following day. Then I noticed one or more "talking heads" or so-called expert commentators repeating and justifying the statements on the news outlets as well as on social media, especially Twitter. In time, I came to believe the tremendous loss in human life from COVID-19 immunization refusal was not an accident but an orchestrated product of a networked political ecosystem of anti-science extremism.

Social Media

Since the beginning, social media has been a leading weapon of choice for anti-vaccine activists. Initially through Facebook, and later Instagram, Twitter, Tik Tok, and other platforms, social media has become a powerful means for anti-vaccine groups to spread disinformation. The Center for Countering Digital Hate (CCDH) in its work to identify a "disinformation dozen" of anti-vaccine activists, based many of its findings on social media use and its impact. Often it is through social media that I first learn I'm a target of anti-vaccine groups. The "OG Villain" moniker, the attacks against me embedded in anti-vaccine books pedaled on Amazon.com, and the dog whistles calling on the patriots to track me down all employ social media as an amplifier.

It stood to reason, therefore, that as vaccination rates stalled across the United States and many refused vaccines, the US surgeon general, Dr. Vivek Murthy, responded by explaining to the American people the role of social media platforms in disseminating misinformation about COVID-19 and COVID-19 vaccines [1]. I was impressed that Dr. Murthy was willing to take on this issue, because prior to the pandemic, the Centers for Disease Control and Prevention (CDC) and other agencies of the Department of Health and Human Services (DHHS) were reluctant to confront most of the elements of the anti-vaccine movement or articulate its dangers to the public. For example, when I wrote an opinion piece titled "How the Antivaxxers Are Winning" in 2017 [2], I received a gentle but firm rebuke from some of the leaders of the CDC, who explained why attempts to publicly mention even the existence of anti-vaccine activities or their adverse impact on public health only served to empower them. As a result, the US government and the DHHS deliberately ignored their ascendency in the hope that the anti-vaccine movement would eventually lose its strength and die a quiet death. While the DHHS maintained a vigorous program of vaccine advocacy and worked hard with state and local health departments to disseminate strong pro-vaccine messages, there was no specific or significant public pushback against those promoting anti-vaccine misinformation. There was no counterweight to the ever-expanding anti-vaccine empire. Thus, the federal government would promote positive messages about the use of the human papillomavirus vaccine (HPV) to prevent cervical and other cancers, but it seldom directly refuted anti-vaccine false assertions that the HPV vaccine caused infertility, named the offending organizations promoting the disinformation, or sought to halt their activities. Such inactions may have created a vacuum that soon filled with anti-vaccine content. During the pandemic, an enabled anti-vaccine movement strengthened even further to become a terrible monster.

To his credit, Dr. Murthy was a pioneer in the DHHS for his willingness not only to bring the anti-vaccine movement into the spotlight but also to announce efforts to begin a counteroffensive. Born in the United Kingdom to immigrant parents from India and raised in Florida, Vivek Murthy was trained at Harvard and Yale Universities, before becoming an attending physician in internal medicine at the prestigious Brigham and Women's Hospital. He quickly acquired a national reputation for organizing efforts to improve healthcare access, ultimately serving as a very young US surgeon general in the Obama administration and later the Biden administration. I grew to admire Dr. Murthy during the pandemic. He made himself accessible, and I have been impressed with his sincerity, intelligence, and work ethic. (As an aside, Vivek and I spoke or communicated by text about Texas Children's Hospital's efforts to develop and administer a low-cost recombinant protein COVID-19 vaccine for children and adults in India. He was supportive and enthusiastic, and I was grateful to have an important champion of this work in the Biden administration.)

On July 15, 2021, Dr. Murthy issued a new advisory and community toolkit [3, 4], calling on the major social media companies such as Facebook, Instagram, YouTube, and Twitter to slow or halt the flow of misinformation. His advisory claimed that two-thirds of unvaccinated adults in the United States believed at least one myth about COVID-19 vaccines [1, 3], including some that I had previously outlined [5, 6]. The myths ranged from the seemingly plausible—that mRNA vaccine production was rushed or that mRNA vaccines were not adequately tested for safety (which was not the case, but it might sound reasonable to many)—to outlandish statements about mRNA creating "genetically modified humans" or somehow magnetizing people. The surgeon general's advisory pointed out how false news stories on social media were far more likely to be spread than accurate health information; it determined that even "brief exposure" to such misinformation was a factor in vaccine hesitancy or refusal [4]. Accordingly, Dr. Murthy called on the social media companies and platforms to alter their computer algorithms so that they would no longer reinforce or amplify negative views about COVID-19 vaccines or other prevention measures [1]. Hoping to limit the spread of health misinformation, he provided a community toolkit to assist families and organizations. He also offered approaches for key stakeholders from the federal government and the health, media, and education sectors to combat misinformation. The mainstream media mostly supported Dr. Murthy. MSNBC's *Morning Joe* broadcast began hitting hard against Facebook and other social media platforms for promoting anti-vaccine rhetoric, as did CNN and others. I also appeared on cable news channels to further those discussions and report in detail about the depth and breadth of the anti-vaccine empire.

The call from the surgeon general and its amplification by news outlets built on previous efforts by the CCDH, which identified the leading anti-vaccine organizations dominating the Internet [7]. CCDH became a major critic of Facebook and Twitter for serving as primary vehicles of this content. Indeed, its landmark report analyzed anti-vaccine content that was posted to Facebook almost 700,000 times over a two-month period, with a more in-depth study of content circulating among anti-vaccine groups that use Facebook as their major platform. In addition, experts such as Renée DiResta at the Stanford Internet Observatory tracked anti-vaccine trends on the Internet, including the dozens of best-selling anti-vaccine books promoted on Amazon.com [8]. Amazon is now a major distributor of anti-vaccine disinformation.

For many years, the social media and e-commerce platforms amplified anti-vaccine disinformation. Especially after writing *Vaccines Did Not Cause Rachel's Autism*, I became a leading target of this weaponized health communication [9]. To anticipate and understand the fallout, I sometimes consulted with both Renée and Imran Ahmed, the founder of CCDH, to seek their advice on how to stem the flow of hate from Facebook groups and the books sold on Amazon. As recent examples, early in the pandemic a prominent anti-vaccine activist branded me as a threat and a "villain" on social media [10, 11] before he was banned from Instagram in 2021. Later, in a 2022 interview published in *Spin* about his Amazon best-selling book, *The Real Anthony Fauci*, he described me thus: "One of Fauci's minions, Peter Hotez [from Baylor College of Medicine], who's a regular on CNN, he's Gates-funded, Fauci-funded—he's never identified as that.... Hotez is a vaccine developer but they don't identify that either. He is dependent on Gates' funding and funding from Fauci. He advocated that we should pass laws making it a felony to criticize Anthony Fauci. And people still look at this guy seriously" [12].

While it is true that I am a vaccine developer—I'm not sure why that should be considered something nefarious—I never said criticizing Dr. Fauci should constitute a felony. At that interview, he also failed to mention that our research group had not received support from the Bill & Melinda Gates Foundation for many years, while the funding for our lab from NIAID-NIH is through a peer-review mechanism that has little or nothing to do with Dr. Fauci's input. Indeed, NIAID-NIH is the world's largest funder of infectious disease research in the United States and globally. Also, the fact that during the pandemic I appeared frequently (unpaid and with no contractual links) not only on CNN but also on MSNBC, NPR, PBS, and many other news outlets (including Fox News for a time) is never communicated. The problem, of course, is that these types of statements are amplified on the Internet through Facebook and other platforms, as are similar statements made about me in his Fauci book, subsequently promoted by Amazon.com. As of this writing, *The Real Anthony Fauci* ranks among the top-rated books on vaccinations at Amazon. Individuals like Renée DiResta and Imran Ahmed have helped me to understand how anti-vaccine activities on social media and e-commerce platforms operate and have suggested potential strategies I can use to counteract some of the hate directed against me.

In 2022, the Stanford Internet Observatory published its update on anti-vaccine activities through a Virality Project conducted in collaboration with the University of Washington, New York University, the Atlantic Council, and other organizations [13]. Among their findings are insights into how the narratives used to discredit COVID-19 vaccines are not new but were adapted from previous false assertions that caught fire and spread rapidly. I have noticed how the mistaken belief that COVID-19 vaccines could cause infertility or were dangerous in pregnancy came right out of the playbook used by anti-vaccine groups to scare women away from taking the HPV vaccine [5, 6]. The Virality Project also highlighted the role of "recurring actors" in promoting anti-vaccine rhetoric, along the lines of CCDH's disinformation dozen, as well as the power of conspiracy theories in accelerating anti-vaccine content. Like the surgeon general's report, the Virality Project reinforced the important role of social media platforms, and how the social media companies must work harder to enforce their own policies and become more transparent in sharing their data on anti-vaccine activities with researchers and government agencies.

In parallel with efforts seeking to understand how the Internet accelerated anti-vaccine disinformation, many national and local vaccine advocacy organizations worked tirelessly during the pandemic to strengthen important pro-COVID-19 vaccine messages. For the most part, these groups, such as the Washington, DC–based Vaccinate Your Family, Voices for Vaccines, the Texas-based Immunization Partnership and Vaxopedia, and many other national or statewide vaccine coalitions were already in operation—in some cases for many years—to prevent the decline in childhood vaccination rates resulting from anti-vaccine groups or access issues. In addition, the International Vaccine Access Center at the Johns Hopkins Bloomberg School of Public Health provides important academic support and evidence-based decision making related to vaccine advocacy, as does the Vaccine Education Center at Children's Hospital of Philadelphia. The Harvard Global Health Institute, together with the Harvard Kennedy School, University of Michigan School of Information, and other partners, constructed a MisinfoR$_x$ toolkit for healthcare providers. Some organizations, such as ThisIsOurShot, specifically formed to promote COVID-19 vaccination rates, while the Rockefeller Foundation launched a new $20 million initiative in 2021 to promote vaccine equity. Within the US government, the National Vaccine Advisory Committee has a long and distinguished history in helping to coordinate vaccination programs, including those for COVID-19 vaccinations, across the DHHS agencies. The CDC created the Vaccines.gov website to provide timely and accurate information about COVID-19 vaccines and their availability. Therefore, in anticipation of resistance to COVID-19 vaccinations, the US government, nonprofit private-public partnerships, and several prominent academic groups lined up to get ready for what we knew might be a tough battle to get every eligible American to accept COVID-19 vaccinations.

During the first year of the pandemic, I had the opportunity to join multiple Zoom calls and meetings with many of these groups and was deeply impressed with their passion and sense of urgency. There is no question that these organizations made important differences in sounding the alarm about the spread of anti-vaccine disinformation on social media, while promoting COVID-19 immunizations, especially in relation to vaccine equity. In time, they sought to refine their messages to appeal directly to vaccine-hesitant communities at risk. However, they also experienced difficulties matching the volume, emotional appeal, fierceness—and at times, cruelty—of the health freedom anti-vaccine lobby. It was also never a fair fight. The pro-vaccine groups worked closely with the academic community to verify their statistics and claims and double-checked them to ensure that they released accurate information to the public. While this was essential and the right thing to do, it also had the downside of slowing down our responses to false assertions. In contrast, the anti-vaccine lobby cared little about the truth or actual science. Instead, it was all about raw emotion and speed. Cornell University psychologist Dr. Valerie Reyna has pointed out how emotional appeals, especially ones that align with core human values, are often far more compelling and convincing than facts [14]. Therefore, the ecosystem of anti-vaccine and anti-science organizations maintained a comparative advantage through its effective, albeit destructive, appeal to emotion and core values. In addition, the pro-vaccine groups remained largely apolitical in their philosophy and approaches during the pandemic. Indeed, such organizations are politically neutral or apolitical by design, especially the ones with nonprofit tax-exempt status or the government entities. However, in this time of COVID-19 they often stumbled when it came to challenging our national partisan divide over vaccine refusal and defiance. The surgeon general's advisory and report did not help us to grasp fully the agenda of the individuals and groups generating anti-vaccine content from the political right—one that resulted in 200,000 "red COVID" deaths. To understand this, we must identify those elements seeking to promote an anti-vaccine and anti-science agenda. Tragically, they stand atop the highest levels of the US government, state governments and the courts, public communications, and in some cases, even academia.

The US Congress

The Conservative Political Action Conference (CPAC) represents one of the largest annual gatherings of conservative political activists. Hosted by the American Conservative Union, since 1974 CPAC has featured and promoted many of the most outspoken and conservative members of the US House of Representatives and Senate, as well as other conservative elected officials and activists. At times in its history, it has become a premier venue for those espousing extreme conservative viewpoints. While the conference has a history of excluding those who promote violence or endanger the public, news outlets have noted that far-right figures now appear at the event, including some of the Oath Keepers and Proud Boys [15]. In 2021, two of these summits were held, the second one taking place in Dallas, Texas, in July. An article in *Esquire*, "They Clapped for Death at CPAC," reported how the Dallas meeting became one of the most notable public expressions of anti-vaccine sentiments and encouragement of vaccine defiance by congressional members and conservative leaders. Especially striking was the enthusiastic applause in response to remarks by Alex Berenson—a former *New York Times* reporter who appears frequently in the right-wing media. As Dave Holmes, the *Esquire* reporter, wrote,

> Alex Berenson is up there crowing about how the Biden administration has missed its goal of getting 90 percent of Americans vaccinated against COVID-19, the virus that brought our world to a standstill from which we are just now starting to come out, the virus that has killed over 4 million people, including 606,000 Americans, and counting. Berenson did not take the responsibility to check his numbers—Biden's goal was 70 percent with at least one dose by July 4, and we're now at roughly 56 percent—but the result is that the audience, upon hearing that a goal of President Biden's has not been met, simply erupts into enthusiastic applause. This roomful of adult human beings puts its hands together for more death. [16]

CPAC also hosted first-term Republican House member Madison Cawthorn, who warned the audience about a Biden White House plan to accelerate vaccine coverage through volunteers knocking on doors to promote COVID-19 immunizations: "Now, they're starting to talk about going door to door to be able to take vaccines to the people. . . . Think about the mechanisms they would have to build to be able to actually execute that massive of a thing . . . [a]nd then think about what those mechanisms could be used for. They could then go door to door and take your guns. They could go door to door and take your Bibles" [17]. Rep. Lauren Boebert (R-Colorado) echoed this sentiment: "Don't come knockin' on my door with your Fauci-ouchie. You leave us the hell alone" [16, 18]. Two days before the start of CPAC, Georgia congresswoman Marjorie Taylor Greene posted on Twitter: "People have a choice, they don't need your medical brown shirts showing up at their door ordering vaccinations" [19], using a reference to Nazi paramilitary groups that helped to propel Hitler's political ascendancy during the 1930s. "You can't force people to be part of the human experiment," she stated.

Each of these congressional members belongs to the House Freedom Caucus, generally considered the most conservative voting bloc among the House Republicans. It was founded in 2015 by proponents of the Republican Tea Party movement (a moniker that has since fallen out of favor), aligned with libertarian and populist sentiments. The group has a reputation for supporting extreme legislation, at times even going up against major conservative platforms. As former House Speaker, Rep. John Boehner once noted: "They can't tell you what they're for. They can tell you everything they're against. They're anarchists. They want total chaos. Tear it all down and start over" [20]. Not surprisingly, the major tenets of health and medical freedom propaganda have found a home among this group.

Rep. Jim Jordan (R-Ohio), who served as the inaugural chairman of the House Freedom Caucus, also openly espoused health freedom anti-vaccine views, especially concerning vaccine mandates, which he refers to as "un-American" [21]. Jordan currently serves as the caucus vice-chair but also as the ranking GOP member of the House Committee on the Judiciary. At the end of 2021 he tweeted: "Hey Fauci, [i]f the booster shots work, why don't they work?" [22]. Tweets such as these set a pattern in which members of the House Freedom Caucus used social media and other mechanisms to discredit the effectiveness of COVID-19 immunizations, or their safety, and to fight vaccination mandates. In July 2021, Rep. Mo Brooks (R-Alabama), another Freedom Caucus member, wrote a letter to President Biden, calling on him to reverse vaccine mandates at Fort Rucker, Alabama. Brooks claimed the vaccines are "experimental," and in a nod to the propaganda of health freedom, he said: "I strongly urge you to respect and protect liberty and freedom and defer COVID-19 response decisions to individual Americans" [23]. Also in July, CNN reported that almost one-half of the House Republicans refused to divulge if they had received a COVID-19 vaccine, with some insisting that they "don't have a responsibility to model behavior to their constituents" [24]. Rep. Chip Roy (R-Texas) offered, "I don't think it's anybody's damn business whether I'm vaccinated or not. . . . This is ridiculous, what we're doing. The American people are fully capable of making an educated decision about whether they want to get the vaccine or not" [24]. Once again, in a nod to health freedom, Rep. Andy Harris (R-Maryland), who also serves as cochair of the GOP Doctors Caucus, affirmed: "Look, we believe in health privacy" [24]. In 2022, Rep. Paul Gosar (R-Arizona), a dentist and also a member of the GOP Doctors Caucus, tweeted, "Ivermectin WORKED as a therapeutic treatment for Covid. Anthony Fauci lied about that fact and hundreds of thousands died because of his malpractice and dishonesty. He belongs in prison" [25]. Many members of the House of Representatives who condemn vaccines may themselves be vaccinated against COVID-19, however. In this way, they prioritize pandering to fringe elements of the GOP over promoting accurate and lifesaving health messages.

Conservative members of the Senate also openly promoted antivaccine views. Sen. Ron Johnson (R-Wisconsin) repeatedly made misleading public statements by twisting and inflating CDC estimates of COVID-19 vaccine side effects. He also questioned the need to immunize young people [26] and even alleged that COVID-19 vaccines could cause HIV/AIDS [27]. At the start of 2022, Sen. Johnson hosted a public forum and panel on vaccine skepticism that was condemned by the Wisconsin medical community [28]. Earlier, the Senate Homeland Security and Governmental Affairs Committee, where Sen. Johnson serves as chairman, invited a physician who openly espouses anti-vaccine views and miraculous cures such as hydroxychloroquine to testify [29]. In the meantime, conservative members of several state legislatures held hearings seeking to discredit vaccines, including one in Ohio where the testimony included claims that COVID-19 vaccines magnetize individuals. One individual held a metal key and bobby pin to her neck and other body parts as alleged proof—although reportedly these metal objects failed to stick [30]. In similar hearings in Louisiana and other states, both Republican lawmakers and those who testified openly promoted anti-vaccine and health freedom propaganda [31]. During the pandemic, Sen. Rand Paul (R-Kentucky) boasted repeatedly that he would not get vaccinated against COVID-19 and made frequent and well-publicized anti-vaccine statements [32].

The GOP congressional members do not speak always with one voice. Texas Republican House members Mike McCaul, Ronny Jackson, and Kevin Brady, together with Sen. John Cornyn from Texas, openly helped to champion the recombinant protein COVID-19 vaccine developed for low- and middle-income countries by the Texas Children's Center for Vaccine Development [33]. Rep. Michael Burgess, a physician from Texas, has also been an ardent supporter of COVID-19 vaccines, as is Senate Minority Leader Mitch McConnell (R-Kentucky), who is also a polio survivor. During home visits to Kentucky in July 2021 McConnell proclaimed, "These shots need to get in everybody's arm as rapidly as possible, or we're going to be back in a situation in the fall that we don't yearn for, that we went through last year." He further said, "I want to encourage everybody to do that and to ignore all of these other voices that are giving demonstrably bad advice" [24]. However, there are about a dozen members of Congress who went out of their way to discredit COVID-19 vaccines and vaccinations. Their major assertions invoke health freedom, claiming that vaccinations are forced on constituents and the American people. Some of these legislators have gone an extra mile to promote disinformation and portray vaccines as dangerous or unnecessary, while others invoke Nazis or offer other extremist rhetoric that encourages vaccine defiance. They have an outsize influence that goes beyond the borders of the states they represent and promotes a national anti-vaccine agenda, causing many Americans to eschew COVID-19 immunizations. In so doing they prioritize the misguided tenets of health freedom propaganda over the lives lost because of vaccine refusal.

Red State Governors and the Courts

Very few of the governors from conservative states have openly questioned the effectiveness or safety of COVID-19 vaccinations. However, in March 2022, Florida became the first state to recommend against COVID-19 vaccinations for children without underlying conditions, going against the recommendations of the Food and Drug Administration and CDC [34]. Later, in October 2022, Florida also recommended against mRNA COVID-19 vaccinations for males 18–39 years of age [35]. Although officially the recommendations came from the Florida Health Department and Dr. Joseph Ladapo, the Florida surgeon general, at least the first one likely arose from anti-vaccine roundtable discussions organized by Republican governor Ron DeSantis. When vaccines first rolled out in January 2021, DeSantis voiced strong support for vaccinating Florida's seniors, but later he began consulting with groups of contrarian physicians and experts who were very public in their opposition to COVID-19 vaccinations and other prevention measures. Dr. Ladapo himself was reported in the press to be linked to the America's Frontline Doctors group, known for its similar views and staunch support of health freedom [36, 37]. The homepage of the organization's website claims that it is "the nation's premier Civil Liberties Organization." Its programs include (1) "Protecting physician independence from government overreach" and (2) "Fighting medical cancel culture and media censorship" [38]. In a nod to the historical role of Dr. Benjamin Rush in helping found the American health freedom movement, he is quoted on its mission statement page. Florida's health freedom zealotry goes beyond vaccinations. Just prior to Dr. Ladapo's announcement, the Florida governor publicly mocked college students who wore masks at an event, claiming that masks were nothing more than "COVID theater" [34]. Ron DeSantis is generally considered a potential GOP front-runner in the 2024 presidential election.

Even though most of the red state governors have tended to avoid questioning the safety or effectiveness of COVID-19 vaccinations, they have been outspoken in their opposition to any sort of vaccine mandates. In 2021, many governors banded together to oppose mandated vaccines for schools, the workplace, or state governments. In a September 2021 evening address, President Joe Biden announced his plan to require federal government and government contractor employees to get vaccinated, in addition to requiring COVID vaccines for healthcare institutions. Then, through an Occupational Safety and Health Administration emergency rule, Biden directed the Department of Labor to mandate vaccines for companies with more than 100 employees. If implemented, such mandates could have helped close the vaccination gap among the 80 million Americans who were eligible for a COVID-19 shot but had refused it.

The response from Republican governors was swift. They criticized the president's remarks and dismissed mandates as unconstitutional and an example of federal government overreach. Such vaccine mandates also violated state sovereignty and personal health freedoms, they claimed [39]. Texas attorney general Ken Paxton simply declared: "Not on my watch!" [40]. Ultimately, the US Supreme Court sided with the governors, voting against the Biden administration's order for the large-employer mandates. The vote was 6–3 and, not surprisingly, strictly along a partisan divide. However, the justices did uphold vaccine mandates for healthcare workers [41]. The stand for health freedom and states' rights versus the federal government became a pyrrhic victory of sorts for the red state governors. The Covid States Project led by the Brookings Institution found that governors who opposed COVID-19 prevention measures or enabled an anti-vaccine agenda (and therefore experienced patient surges that overwhelmed hospitals and intensive care units) also suffered significant drops in their public approval ratings [42]. The legacy of these governors might become one of prioritizing a small but vocal health freedom lobby over commonsense public health interventions.

The Supreme Court's rebuke of the Biden administration's efforts to enact COVID-19 vaccine mandates was followed by additional federal or state court rulings to thwart public health interventions. In 2022, a US district judge in Florida, Kathryn Kimball Mizelle, reversed federal mask mandates for domestic air travel and other forms of public transportation [43]. The ruling occurred just as the dangerous BA.2.12 subvariant of Omicron, the most transmissible SARS-2 coronavirus seen up until then, was accelerating in New England and the mid-Atlantic states, and was in response to a suit brought about by the Health Freedom Defense Fund [44], which also seeks to strike down vaccine mandates [45]. Judge Mizelle is one of the youngest district court judges, appointed by President Trump two months before he lost his 2020 presidential bid. This sets a dangerous precedent in which federal judges without qualifications in medicine or public health can reverse the decisions made by experts, including those recommended or implemented by the CDC to protect the US population. Another example: In the summer of 2022, Trump-appointed US District Judge Terry Doughty ruled that two GOP attorney generals from Missouri and Louisiana could proceed with the discovery phase of a lawsuit accusing Drs. Anthony Fauci and Vivek Murthy, along with other Biden administration officials, of colluding to suppress COVID-19 "freedom of speech" [46]. While the Biden administration can use a system of appeals to attempt to limit the damage caused by such rulings, such mechanisms are often slow and unwieldy. They also fail to halt the political damage that such court battles could engender.

Conservative News Outlets

While elected officials at the federal and state levels did much to push a health freedom agenda, they were matched with equal fervor by the major conservative news outlets. The largest, Fox News, is king, and as someone who appeared frequently on CNN, MSNBC, BBC, and other news channels throughout the pandemic, I became familiar with many of the network's anchors and presenters. Early in the pandemic, I appeared several times on Fox News nighttime programming and even did interviews with two of its most-watched anchors—Tucker Carlson and Sean Hannity. However, as I began to call out the Trump White House in 2020 for spreading disinformation about the pandemic, including false claims that downplayed the severity of COVID-19, comparing it to the flu, and aggressively promoting hydroxychloroquine as some sort of magic cure, I was eventually cut out of the evening hours. While I still appeared from time to time on Fox News, it was only during the daytime. At night I was replaced by a group of commentators who supported the views endorsed by the Trump White House.

In 2021, Fox News, one of the most-viewed news organizations, became a powerful and influential promoter of health freedom and anti-vaccine viewpoints [47–51]. In the summer of 2021, just after COVID-19 vaccines became widely available, a watchdog group that tracks conservative media known as Media Matters for America (Media Matters) recorded 840 Fox News "negative claims" that undermined efforts by the Biden administration to vaccinate the country [47–49]. Specifically, according to Media Matters, the network "argued that immunization efforts were coercive or violated freedom or choice 554 times (66% of all claims), that vaccines were dangerous or unnecessary 375 times (45%), and that Democratic efforts to vaccinate the populace were a cynical political ploy 95 times (11%)" [49]. Media Matters is a nonprofit organization but is known as "left-leaning" in its political orientation and donor support. It was founded in 2004 by David Brock, and according to a 2008 *New York Times* article, uses its funding "to monitor and transcribe nearly every word not only on network and cable news but also on nationally syndicated talk radio and, lately, local radio" [52]. The organization's 2021 report stated: "In a six-week period from June 28 through August 8, Media Matters found that nearly 60% of the network's vaccine segments included claims undermining or downplaying vaccinations" [49], and this was also reported elsewhere [47].

While some of the daytime news hosts such as Harris Faulkner—a Fox news anchor who previously interviewed me on several occasions (and someone I thought was a straight shooter and very reasonable)—participated in a pro-vaccine public service announcement, these efforts were undermined by her more influential nighttime counterparts. Media Matters cited Tucker Carlson and Laura Ingraham, two of the most-viewed Fox anchors, as major disseminators of anti-vaccine disinformation [48, 49]. Media Matters noted as well that in the six-week period beginning at the end of June until the beginning of August 2021, almost one-half of the Fox News vaccine segments either alleged Biden White House overreach or focused on "coercive" violations of health freedom [49]. The organization also found that one-third of the segments claimed that COVID-19 vaccines were either dangerous or unnecessary [49]. The nighttime anchors were especially egregious, according to the Media Matters report: "*The Ingraham Angle* aired negative claims in 98% of its vaccine segments, *Hannity* was second with 91%, and *Tucker Carlson Tonight* and *Fox News Primetime* followed with 90% and 89%, respectively" [49]. The report summarized the major anti-vaccine sentiments from Fox News anchors, and to me they closely resemble the health freedom propaganda. Specifically, Media Matters listed the following topics as regular elements of Fox News broadcasts in summer 2021, as the Delta wave unfolded in America [49]:

- Vaccines are unnecessary or dangerous.
- Immunization efforts are coercive, represent government overreach, or violate personal freedom or choice.
- Vaccination efforts are a cynical political ploy by Democrats.

Along those lines, on August 9, 2021, Carlson "likened immunization requirements to forced sterilization and lobotomies" [50], while comparing vaccine passports to "Jim Crow" laws, according to CNN [51]. CNN on its website has also called attention to Fox News's COVID-19 vaccine self-reporting policies that in some respects resemble vaccine passports, calling out the organization as "brazenly hypocritical" [51]. In the meantime, the Fox News attacks against COVID-19 vaccinations continued into 2022, with an analysis by the *Washington Post* revealing how the network used its broadcasts and social media feeds on Twitter to highlight those who have fallen ill after receiving COVID-19 vaccination—whether or not the two events were in fact actually related [53].

In a second independent analysis published in the peer-reviewed journal *Scientific Reports* (in the *Nature* family of journals), the Center for Law and Economics at Switzerland's ETH Zurich (also known as the Swiss Federal Institute of Technology), which is sometimes considered Europe's equivalent of the Massachusetts Institute of Technology, found that starting in May 2021 "viewership of Fox News Channel has been associated with lower local vaccination rates" [54, 55]. It also confirmed that "Fox News increased reported vaccine hesitancy in local survey responses," whereas "two major television networks, CNN and MSNBC, have no effect" [54]. The ETH Zurich group also found that vaccine hesitancy was "not due to the general consumption of cable news" [54].

Fox News disinformation is not restricted to vaccines and vaccinations. An extensive *New York Times* analysis of more than 1,000 Tucker Carlson broadcasts conducted in 2022 finds them heavily laden with conspiracy theories and far-right extremist viewpoints, noting Carlson's privileged access to the Fox organization's chairman, Rupert Murdoch [56]. In addition, Murdoch's son Lachlan is the CEO of Fox Corporation, and according to Media Matters in 2021, "Fox Corp. CEO Lachlan Murdoch has publicly endorsed a debunked conspiracy theory spread by Fox News prime time host Tucker Carlson, who has claimed that thousands of people died in connection with the COVID-19 vaccines" [57]. To be sure, Fox News was not the only conservative news site promoting these opinions, but its unparalleled and extensive reach into America's homes at night greatly amplified the anti-vaccine and health freedom viewpoints of certain elected officials and public intellectuals. According to the Nielsen ratings for television viewing, *Tucker Carlson Tonight* was the highest-rated cable news show in 2021, with 3.21 million viewers, followed closely by third-ranked *Hannity* (2.87 million), and *The Ingraham Angle* at fifth with 2.27 million [58].

In addition to serving as chairman of the Fox Corporation, Rupert Murdoch, the Australian-born media tycoon (and multibillionaire), is also executive chairman of News Corp, which oversees the *Wall Street Journal*, *New York Post*, and *The Sun* in the United Kingdom, among other media holdings. The *Wall Street Journal* provides an interesting case study. While the newspaper's mainstream reporting seems mostly uncontaminated by anti-vaccine sentiments, the opposite is true of its opinion section, where several questionable pieces have appeared. These include an article coauthored by Dr. Ladapo shortly before he was appointed Florida's surgeon general, "Are Covid Vaccines Riskier than Advertised?" [59], another authored by a medical school professor with the title "Is the Coronavirus as Deadly as They Say?" [60], and a third that enthusiastically boasted "We'll Have Herd Immunity by April [2021]" [61] in the months before America's devastating Delta variant wave. In 2021, Joseph Azam, who was a former senior vice president at News Corp, denounced Fox News for its role in "turbo-charging" (according to the *Sydney Morning Herald*) vaccine hesitancy and resistance among conservatives in the United States. In an interview with that newspaper, Azam said, "I think Fox has been almost single-handedly responsible for the politicisation of public health in the US and the creation of vaccine hesitancy in a significant portion of the population.... It's been tremendously damaging" [62].

He was not alone. In 2020, James Murdoch, Rupert Murdoch's younger son, resigned from the News Corp board, later telling Maureen Dowd in a *New York Times* interview, "I think at great news organizations, the mission really should be to introduce fact to disperse doubt—

not to sow doubt, to obscure fact, if you will" [63]. Shortly afterward, in December 2020, Rupert Murdoch became one of the first individuals to receive a COVID-19 vaccination [64].

The Contrarian Intellectuals: A Toxic Alliance

Piling on to this anti-vaccine assembly of elected officials and conservative news outlets is a cadre of public intellectuals who appear on Fox News and other venues. These include physicians and even professors at major academic health centers. The impact of America's Frontline Doctors was mentioned above, but in addition, Media Matters found that in the six-week period highlighted above, a prominent Johns Hopkins University medical school physician made 45 claims "undercutting" vaccination, as have others from Stanford University and elsewhere [49]. Some of the intellectuals promoting anti-vaccine views are now connected to newly established libertarian think tanks, as well as prominent US universities [65–70]. My colleague, Dr. David Gorski, a cancer surgeon at the Detroit Medical Center and professor at Wayne State University has written extensively (under a publicly disclosed pseudonym known as "Orac") about the Brownstone Institute, a Texas-based libertarian think tank that was founded during the pandemic and has become a sort of intellectual home for contrarian intellectuals opposed to masks and other COVID-19 preventions, including vaccine mandates [67, 70]. Having credentialed professors and other academics speak out against vaccines or downplay the severity of COVID-19 provides academic cover for the far-right or even extremist views promoted by some members of Congress, conservative governors, and outlets like Fox News. Oftentimes, the arguments of the contrarian intellectuals are extremely clever, using real facts woven together in devious ways to spin false narratives about the ineffectiveness or harmful outcomes of COVID-19 vaccinations and other prevention measures. Some efforts have been made to reveal the financial interests backing these conservative or libertarian intellectuals and organizations. A Google search reveals some articles alleging ties to conservative political donors, although in my view such claims are still mostly speculative and unproven.

Although not a physician, Alex Berenson, who spoke out against vaccines at the July 2021 CPAC summit, is an occasional guest on Fox News [16, 49, 71]. On the January 25, 2022, *Tucker Carlson Tonight* broadcast, Berenson told Carlson, "The mRNA Covid vaccines need to be withdrawn from the market. . . . No one should get them. No one should get boosted. No one should get double boosted. They are a dangerous and ineffective product at this point" [72]. Previously, Mr. Berenson was suspended from Twitter (although later reinstated [73]) and featured in an April 2021 article in the *Atlantic* titled "The Pandemic's Wrongest Man" [74]. Berenson and other prominent vaccine skeptics, including those connected to the "Intellectual Dark Web," who challenge liberal ideologies while in some cases openly espousing anti-vaccine viewpoints, have appeared on the *Joe Rogan Experience* [75], one of the most popular podcasts around the world, with more than 100 million downloads per month. In fairness to Joe Rogan, I have also appeared on his podcast twice, as has my esteemed colleague, Dr. Michael Osterholm, but the damage done by such wide dissemination of anti-vaccine views may have contributed to vaccine hesitancy in America. I am concerned about the powerful negative impact generated by prominent intellectuals, physicians, and scientists who speak out against vaccines and COVID-19 prevention. This constitutes an important cog in the anti-vaccine and anti-science machine that has cost so many American lives.

The right-leaning elected officials and the courts, conservative news outlets, and contrarian intellectuals combine to create a dangerous alliance that now dominates the news media and Internet. In 2021, these forces converged with some of the anti-vaccine activists belonging to the disinformation dozen. For example, toward the end of 2021, NPR reported on how pro-Trump Republicans and anti-vaccine activists met at a joint conference held in Nashville, Tennessee, in October [76]. This alliance of individuals and groups with anti-vaccine and other anti-science agendas based on health freedom propaganda are now firmly linked with those espousing conservative and at times, far-right, or even extremist views. Shortly after the January 2022 anti-vaccine and anti-science protests in Washington, DC, *Time* ran an article titled "How the Anti-vax Movement Is Taking Over the Right" [77], although in my opinion it could easily have read, How the Right Is Taking Over the Anti-vax Movement. In either case, the story paints a dark picture of a dangerous movement that now includes the Proud Boys and other extremist elements. Along those lines, it is interesting to note that prominent anti-vaccine activists were arrested for their participation in the January 6 storming of the US Capitol [78–81]. Together this anti-vaccine political ecosystem on the right has been highly persuasive in encouraging thousands of Republicans to avoid COVID-19 vaccinations, especially after the vaccines became widely available in the middle of 2021.

This media and political empire is causing unprecedented losses of human life. The pervasive role of disinformation from this segment of society has not gone unnoticed by the Biden administration. In 2021, they proposed forming a new disinformation advisory board through the Department of Homeland Security to begin tackling issues related to not only COVID-19 prevention but also the 2020 US presidential election and other key issues [81]. However, these efforts met with significant opposition from the Senate GOP. As one former intelligence official in Homeland Security pointed out, "You can't even use the word 'disinformation' today without it having a political connotation" [81]. For now, the advisory board has been tabled, and the anti-science political ecosystem continues largely unchallenged.

> He's [Hotez is] a misinformation machine constantly spewing insanity.
>
> —Tucker Carlson, Fox News

> I have threats upon my life, harassment of my family and my children with obscene phone calls because people are lying about me.
>
> —Anthony Fauci, MD

5 | A Tough Time to Be a Scientist

The attacks have taken a personal toll. Over the past 20 years, since I first began writing or speaking about the anti-vaccine movement and debunking its assertions about autism, I have been a target for anti-vaccine groups. They initially claimed that the measles-mumps-rubella (MMR) vaccine itself caused autism but later shifted their position to insist that the disorder must be triggered by administering vaccines too closely together, or by the thimerosal preservative or alum salts in some vaccines. As quickly as I, and other scientists, debunked their falsehoods, the anti-vaccine activists generated new ones. I explained the evidence showing there is no connection with autism and detailed the genetic and epigenetic bases of autism or their associated intellectual disabilities. I even wrote about my daughter Rachel's autism gene. For the most part, anti-vaccine individuals or organizations sought to discredit my statements or dismiss them, making (knowingly) false claims that I belonged to a pharmaceutical industry cabal or secretly took pharma funds. In fact, since 2000 I have led or co-led an academic-based vaccine research group based in Texas, where we develop new, low-cost vaccines for resource-poor countries. We sometimes refer to them as the "antipoverty vaccines" because of their societal impact [1]. Most recently, our laboratory research group developed a prototype COVID-19 vaccine technology leading to CORBEVAX, which is produced by Biological E, one of India's major vaccine companies [2]. In a major pediatric campaign during the first half of 2022, more than 75 million doses of this vaccine were administered to schoolchildren in India, and the vaccine has been since approved in Botswana. We also developed a second and related vaccine technology that helped Bio Farma, a vaccine manufacturer in Indonesia, produce IndoVac.

None of this, of course, seemed to matter much to anti-vaccine groups. My old friend and colleague Dr. Paul Offit from the Children's Hospital of Philadelphia (where he led the development of a vaccine for rotavirus infection) and I were labeled "OG Villains." I had to look this up and was not happy when I learned that OG stands for "original gangster"; apparently, we are considered gangsters for our historical and staunch defense of childhood vaccinations [3, 4]. Prior to the pandemic we were featured in a meme that circulated widely on the Internet. It was drawn or made to resemble the gangsta rap group N.W.A. and the *Straight Outta Compton* movie poster. However, we were depicted as OG pharma shills belonging to the "the world's most toxic group" with a new *Straight Outta Merck* title, presumably referring to the company that produces the MMR and human papilloma virus vaccines.

Even prior to the pandemic, it was not easy to defend against the OG Villain title. For example, in November 2019, just as COVID-19 was possibly emerging in central China, I was speaking at an infectious disease conference in New York City about our vaccines for parasitic and poverty-related neglected diseases. Typically, when I'm invited to speak at an event or to give a grand rounds lecture, I try to arrive the evening before to collect my thoughts and have a quiet dinner. I put a lot of passion and emotion into my lectures while providing detailed content—

meaning I take my lectures very seriously and give them my full attention and effort. So, the downtime the evening before is always really welcome. Unfortunately, on this occasion no such calm preparation was possible. Returning from dinner, I was accosted at the hotel entrance by a couple with anti-vaccine beliefs. This was not the first time something like this happened. It usually proceeds along the following lines: one person asks me provocative questions about vaccine safety or parents who refuse to vaccinate their children while the other records my responses on a cell phone. Within hours or days, an edited recording typically shows up as a YouTube video along with their commentary. After that, anti-vaccine groups can adapt the video for their websites. For this event in Manhattan, the recording was not the end. On the following day, anti-vaccine activists demonstrated in front of the hotel entrance to protest my lecture and those of other conference participants. To escape so as to make my travel connections safely and unnoticed, hotel security whisked me out through a secretive back exit into a waiting car.

By this time, the anti-vaccine movement was especially strong in Texas where health freedom propaganda had first gained traction a few years before. As a vaccine scientist working in the state, and as someone with a growing national reputation for opposing nonmedical vaccine exemptions for schools [5], I became an increasingly visible target [6]. The first reason was my staunch defense of vaccines and public advocacy to uncouple them from the causes of autism or neurodevelopmental disorders. However, the second reason centered on politics. By opposing what was then a new health freedom trend, I unknowingly became a person regarded by the far-right as suspicious. In 2019, a Texas GOP legislator, Rep. Jonathan Stickland, directly condemned me in a couple of open messages on Twitter:

> You are bought and paid for by the biggest special interest in politics. Do our state a favor and mind your own business. Parental rights mean more to us than your self-enriching 'science'. #txlege. [7]

Make the case for your sorcery to consumers on your own dime. Like every other business. Quit using the heavy hand of government to make your business profitable through mandates and immunity. It's disgusting. [8]

While the pharma shill accusations were nothing new, I found it truly shocking that an elected member of the Texas state legislature would launch an unprovoked public attack against a private citizen, especially against a scientist working at the Texas Medical Center to develop lifesaving vaccines for the world's poor. It was also bizarre and hurtful that he attempted to reduce my life's work to "sorcery." This was something I could not have imagined during my extensive doctoral and research training decades before. I was not the only one aghast at this behavior. Rep. Stickland's vituperative denunciations of a scientist in his own state made national headlines in the *New York Times*, *Washington Post*, CNN, and elsewhere [9–11]. At that time, I did not realize that his misbehavior was just a warm-up for something more ominous.

Not Only the Science but Also the Individual Scientists

As the COVID-19 pandemic unfolded, the attacks against me continued. In response to my article in *Nature* (in the spring of 2021) that highlighted the risk of a globalizing anti-vaccine movement [12], a well-known anti-vaccine activist and promoter of conspiracies compared me to Hitler and Stalin in his publication [13, 14]. He then proceeded to list my contact information, including my e-mail address and office telephone number, an intimidation tactic known as doxing. A torrent of threats via e-mail followed, with several calling for my torture or public execution. One suggested that my death should be viewed on television as I received 2,000 doses of "Satan-Vax" (fig. 5.1), while others wished that I would perish via more conventional means, such as the hangings that followed the trials at Nuremberg at the end of World War II. There were also intimidating phone calls and voicemail messages left at my office.

The practice of comparing those responsible for developing, distributing, or promoting COVID-19 vaccines to Nazis became a prominent theme during the pandemic, especially in the months following the release of the Pfizer-BioNTech and Moderna mRNA vaccines in 2021. Another e-mail compared me to a "living Mengele," referring to the infamous Dr. Josef Mengele, an SS officer who conducted human experimentation at the Auschwitz concentration camp (fig. 5.2). He was known as the "angel of death" and was personally responsible for countless murders, while many more victims were hideously tortured. Mengele ranked among the most vicious sadists and virulent anti-

Figure 5.1. E-mail sent to me on May 9, 2021. The dark bars here and in the next messages redact the possible names of the senders, as well as my e-mail address.

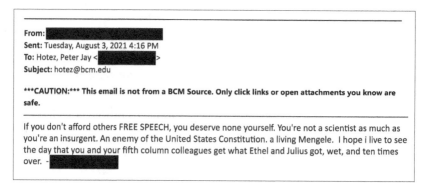

Figure 5.2. E-mail sent to me on August 3, 2021.

Semites in the Nazi killing machine. The last sentence makes an allusion to Ethel and Julius Rosenberg, two Jews executed in 1953 by the US government on espionage charges despite numerous international appeals, and wishes me the same fate.

Additional Mengele analogies also appeared on Twitter. At some level, anti-vaccine groups and individuals wished to make the case that mRNA vaccines are experimental and that the pharma companies were treating the US population as guinea pigs. As a vaccine scientist and someone who endorsed mRNA vaccines during my media interviews on high-profile cable networks [15], the public saw me as a thought-leader for these practices. The vicious threats that I received did at times take a toll on my mental health. On many occasions I lost my concentration at work or woke up in the middle of the night because I was so upset by these unfounded accusations. In addition to sadness, my other emotion was righteous indignation. After all, I obtained my MD and PhD and worked all my life to develop lifesaving interventions for diseases of the poor. Now a segment of American society sought my public execution in a manner befitting a Nazi doctor. The fact that I am Jewish and had family members suffer in the Holocaust made this period especially demoralizing. Increasingly, I began to notice a connection between anti-science and anti-Semitism [16]. I was targeted in this manner in part because I am a Jewish scientist, and many elements of the far-right embrace attacks on and harassment of the Jewish people [17].

However, it was also clear that I was not the only American scientist called "Nazi" and "Dr. Mengele." Dr. Fauci was also under attack, despite the fact that since 1984, he had advised six US presidents about major infectious disease threats facing the nation, beginning with HIV/AIDS, and continuing with H1N1 pandemic influenza, Ebola virus infection, and most recently COVID-19. Tony was even awarded the Presidential Medal of Freedom and the Lasker-Bloomberg Prize for his public service. Dr. Fauci has also been an important mentor and friend to me personally. We both obtained our MD degrees from Weill Cornell Medical College (although many years apart), and I have often sought his advice and counsel, especially concerning important career shifts, such as my relocation to the Texas Medical Center. Tony has also provided important input regarding my battle with anti-vaccine forces. As the COVID-19 pandemic unfolded, he endured the same Nazi and Dr. Mengele accusations, including one made by a host of Fox Nation, a streaming service of Fox News [18, 19]. Via Twitter, the Auschwitz-Birkenau Memorial and Museum appropriately condemned these remarks: "Exploiting the tragedy of people who became victims of criminal pseudo-medical experiments in Auschwitz in a debate about vaccines, pandemic[s] and people who fight for saving human lives is shameful. It is disrespectful to victims & a sad symptom of moral and intellectual decline" [20]. I am thankful that the Auschwitz Museum also rose to my defense after similar Nazi-related comparisons and threats.

Unfortunately, the Fox News threats did not stop there. Multiple conservative news outlets have criticized Dr. Fauci for a potential role the NIAID-NIH had in supporting a specific type of virologic research to enhance the virulence of virus pathogens. With respect to coronavirus research, this includes unfair attacks claiming he had a role in generating the SARS-2 coronavirus that caused the pandemic. The far-right made unfounded claims that SARS-2 originated in a laboratory through research activities supported by NIAID-NIH. In a gathering of conservatives, one Fox News host used violent language in referring to how journalists should approach Dr. Fauci about these allegations: "Now you go in for the kill shot. The kill shot with an ambush, deadly, because he doesn't see it coming.... This is when you say, 'Dr. Fauci, you funded risky research at a sloppy Chinese lab, the same lab that sprung this pandemic on the world. You know why people don't trust you, don't you?' Boom, he is dead. He is dead. He's done" [21, 22]. According to the *New York Times*, Fox News regularly refers to "Lord Fauci" as someone who is attempting to take basic freedoms away from Americans, while the *Washington Post* reports that Tucker Carlson has made statements claiming Dr. Fauci was "the guy who created Covid" or that he is someone who creates or disseminates "authoritarian germ hysteria" [21, 23]. In an article in *Science* magazine published in the summer of 2022, the distinguished science journalist Jon Cohen wrote an article with the straightforward title "Almost Everything Tucker Carlson Said about Anthony Fauci This Week Was Misleading or False" [24]. Also that month, Florida governor Ron DeSantis said about Dr. Fauci: "I'm sick of seeing him! I know he says he's gonna retire. Someone needs to grab that little elf and chuck him across the Potomac [River]" [25].

The Newest Enemies of the State

The attacks against Dr. Fauci and other US scientists in the fall of 2021 did not end with the conservative news outlets. Next, elected members of Congress and a sitting GOP governor went on the attack. Florida governor Ron DeSantis, considered a party leader, labeled Dr. Fauci a "Covid authoritarian," while Texas senator Ted Cruz, in a Fox News interview, said he was "the most dangerous bureaucrat in the history of the country" [21]. Also on Fox News, Ohio representative and Republican Jim Jordan said, "The American people want safe streets, affordable gas and freedom, and the Biden administration has given us record crime, record inflation and Dr. Fauci" [21]. All these statements followed on the heels of an earlier effort by Rep. Marjorie Taylor Greene (R-Georgia) to introduce the Fire Fauci Act to halt his federal salary and force his retirement [21]. Kentucky senator Rand Paul added fuel to the demands to oust Tony through his unfounded accusations and public confrontations with Dr. Fauci at Senate hearings [26].

The threats and accusations against Tony by members of Congress may have been the trigger for additional death threats or calls for violence. While it is difficult to connect all the dots, the fact that congressional leaders of the far-right and other elected officials specifically targeted Dr. Fauci, while Fox News and other news outlets amplified this disinformation, may have led to further actions. Possibly this reckless spread of disinformation about Tony may have acted as dog whistles for the unwell to take matters into their own hands. At the end of 2021, a California man was arrested in Iowa after a traffic stop revealed he carried an AR-15-type assault rifle and intended to drive to Washington, DC, to assassinate Dr. Fauci, among others [27]. In 2022, a GOP candidate for Congress, then the chairman of the Oklahoma Republican Party, stated during a campaign speech that "we should try Anthony Fauci and put him in front of a firing squad" [28], while a man in West Virginia was sentenced to three years in prison for sending threatening e-mails to Dr. Fauci, one of which stated, "I will slaughter your entire family. You will pay with your children's blood for your crimes" [29]. The defendant admitted in court that his goal was to intimidate Dr. Fauci and others in the US government who had discussed the testing and prevention of COVID-19 [29]. Finally, the conspiracy website Real Raw News made claims that Dr. Fauci had been arrested by the US military and subsequently hanged at Guantanamo [30]. The piece is especially sickening, as this twisted fantasy describes Dr. Fauci's purported execution by hanging in gruesome detail.

In an article in the *Daily Beast* titled "The Unique Terror of Being a COVID Scientist after Jan. 6," I wrote that such threats constitute an enlargement of the attacks against science by taking specific aim—in some cases literally—against American biomedical scientists [31]. In the case of Dr. Fauci, the grievances put forward by members of Congress and other elected officials from the far-right claimed that he represented a major threat to health freedom. They asserted he did this to wield power and exert authority over masks, social distancing, and ultimately, vaccine mandates. Among the most egregious accusations were their claims that Dr. Fauci led efforts to engineer with deliberate or nefarious intent the SARS-CoV2 virus, the cause of COVID-19, through NIAID-NIH funding of joint Sino-US coronavirus research collaborations [31]. They made unsubstantiated claims that Chinese scientists working at the Wuhan Institute of Virology conducted research to make coronaviruses especially deadly—sometimes referred to as gain of function (GOF) research—before the virus leaked from the Wuhan laboratory to ignite the pandemic. Far-right groups further claimed that GOF research depended on US government NIAID-NIH funds, and that a US organization known as the EcoHealth Alliance, based in New York City and headed by Dr. Peter Daszak, served as an intermediary.

For me, what was especially shocking were efforts by House Republicans and other elected officials, journalists, and even some well-established American scientists (although mostly from fields outside of virology) to pile on against Tony and Peter, harassing them with endless Freedom of Information Act requests and demanding a seemingly unending stream of investigations [32]. Such requests ignored the reality that there was compelling evidence for the natural origins of COVID-19 emerging from exotic animal wet markets [33, 34], or the reality that many viruses similar to SARS-CoV2 were already present in bats across East Asia [35]. The conspiracy theorists also made exaggerated claims about finding a signature amino acid sequence known as a furin cleavage site, claiming it represents a telltale or "smoking gun" sign of human manipulation. However, such assertions ignored the findings that furin cleavage sites exist in just about every subfamily of coronaviruses [36, 37] even though SARS-CoV2 was the first example in its own subfamily. The Tulane University virologist Dr. Robert Garry has written eloquently in the *Proceedings of the National Academy of Sciences USA* on why the GOF or lab leak theories for COVID-19 origins do not stand up to scientific rigor [38, 39], whereas there is overwhelming evidence for the natural origins of COVID-19 jumping from bats or other animal intermediate hosts to humans [33, 34, 40, 41]. Especially troubling is the fact that such realities did not prevent members of Congress from making calls to withhold federal funding for the EcoHealth Alliance [42]. Late in the summer of 2022, Sen. Paul, Rep. Jordan, and other prominent Republican congressmen announced their interest in launching extensive investigations against Dr. Fauci and others in the Biden administration, including threats to subpoena them, if both houses of Congress were to fall under GOP control [43].

This was by no means the first time that elected officials abused their power to unfairly target biomedical scientists. For a decade, Rep. John Dingell (D-Michigan) relentlessly pursued Nobel laureate Dr. David Baltimore and his associate Dr. Thereza Imanishi-Kari for alleged scientific fraud [44], before their exoneration by a Department of Health and Human Services scientific review panel. According to one report, "Dingell, who headed the House Energy and Commerce Committee, took up the issue, treating scientists with the same forcefulness and disdain he had used with overcharging military contractors" [45]. Similarly, while he was Connecticut attorney general, Sen. Richard Blumenthal (D-Connecticut) aggressively pursued the Infectious Disease Society of America for its guidelines that protected patients from unnecessary use of antibiotics in the treatment of Lyme disease [46]. In both cases, these elected officials targeted science and scientists for political gain or for reasons that went outside the bounds of any serious scientific inquiry. However, for the most part these were one-off engagements and did not reflect a systematic and organized campaign such as the one we are seeing today by far-right elements of the US government.

As the attacks against both Tony Fauci and Peter Daszak unfolded in 2021, I reached out to them and offered my support. I did so because of my personal experiences, realizing how awful it can be to be attacked unjustly in public, especially when the only crime supposedly perpetrated was using scientific knowledge to save American lives. As in my case, the attacks against Drs. Fauci and Daszak went beyond electronic threats via e-mail or social media. For instance, in a heated exchange at a Senate hearing in January 2022, Dr. Fauci held up printouts of the "Fire Fauci" tab on Sen. Paul's reelection campaign website [47]. Dr. Fauci also added, "I have threats upon my life, harassment of my family and my children with obscene phone calls because people are lying about me" [48]. In addition, he discussed the arrest of the California man who planned an assassination attempt. Similarly, in multiple phone and Zoom calls with Peter Daszak I learned about similar threats against both Peter and the EcoHealth Alliance. He was accosted at his home, and he has had to ask for help from law enforcement because of death threats via e-mail and on social media; a white powder was even sent to his home. Dr. Fauci and his family members have required federal protection. I have not had too much time to speak with Tony owing to his many COVID-19 government responsibilities, but it is clear the threats against him from the far-right have been a source of great sadness and disappointment. In the meantime, a study and analysis conducted by *Nature* reports that threats against biomedical scientists are on the rise and are not just confined to the United States [49]. They include multiple prominent UK, European, and Australian scientists who study COVID-19 or who speak in the media. Another study of almost 10,000 researchers discovered that of the more than 500 who responded, almost 40% experienced an insult or death threat via social media, e-mail, or phone [50]. The study mentions that in some cases there were also direct physical confrontations. Scientists who openly discussed their pro-vaccine views or dismissal of GOF as accounting for COVID origins, or those discrediting ivermectin and hydroxychloroquine as effective therapeutics were targets. Essentially these are scientists who refute health freedom propaganda. Many of them experience mental health issues, including anxiety, lost productivity, and fear of loss of reputation, loss of employment, or safety [50]. However, the examples of Drs. Fauci and Daszak point to something especially insidious, namely, attacks and threats that are either tacitly or overtly endorsed by US elected officials.

Under Attack

Following the combined onslaughts from elected officials and conservative news outlets against Drs. Fauci and Daszak, the public attacks against me accelerated. On June 22, 2021, on the Fox News nighttime broadcast of *The Ingraham Angle*, Laura Ingraham interviewed Florida governor Ron DeSantis about me after showing a video clip in which I predicted an imminent wave of COVID-19 across the southern states. Here is the transcript:

> DR. PETER HOTEZ, PROFESSOR AND DEAN OF TROPICAL MEDICINE, BAYLOR COLLEGE OF MEDICINE: Travel, and especially over the July 4th holiday, that could be a big issue. We saw that summer surge, and that was pretty awful in a belt that went all the way from Arizona through New Mexico, Texas, across the Gulf Coast into Florida. We just have to assume that mother nature is telling us, this is going to happen again.
>
> (END VIDEO CLIP)
>
> INGRAHAM: That's the infamous Dr. Peter Hotez, Governor. They just can't let the pandemic go [*sic*]. At some point, they're going to have to break the addiction.
>
> DESANTIS: And it's like some of these people get put out there all the time when they have been dead wrong over the last year. For example, a lot of these experts criticized Florida for getting our kids back into school in August. They said, this will be two or three weeks, everyone is going to get sick, all the schools are going to have to shut down again. That just never

happened. I think schools were probably one of the places that had the fewest amount of infections of anywhere else in our society. So you are wrong on these really, really big issues that impacted millions of people, and you are still out there parroting stuff. [51]

Of course, it turned out I was right and had correctly predicted a horrific COVID-19 wave that inundated Florida in the summer of 2021, causing many hospitalizations and deaths. Just three weeks later, the *Daily Kos* wrote an article titled "Ron DeSantis Fed COVID Crow by Doctor He'd Ridiculed on Fox News," opening with the following statement: "Oh, this is a rich, gooey, chocolatey, delicious slice of schadenfreude a la choad. Florida Gov. Ron DeSantis, who keeps saying his handling of the coronavirus was totes rad yo, even though it so totally was not, may have to wipe the smug look off his face as COVID cases in his state continue to dangerously escalate" [52]. However, for me there was no joy or real satisfaction—I was horrified at the needless loss of human life in Florida because my warnings, along with those of other US public health officials, may have been ignored. In addition, it was becoming clear that negative comments about me on a widely viewed conservative news outlet, especially from one of the nighttime Fox News anchors together with one of the leaders of the GOP, inspired further attacks on the Internet.

In time, I became a new favorite target for the far-right. As a guest on Steve Bannon's *War Room* podcast for Real America's Voice, Rep. Greene referred to me as "Mr. Bowtie, who calls himself a scientist." Previously I had criticized her public anti–COVID-19 vaccine statements as contributing to the disinformation leading to needless American deaths. Presumably, she was unaware of my role in developing a COVID-19 vaccine. She continued, "He thinks he's the authority of truth. . . . Here's the situation, scientists have been wrong over and over and over since the beginning of time. So just because he's a scientist doesn't mean he's right" [53]. Rep. Greene also used the opportunity to spread additional disinformation about COVID-19 vaccines.

We were clearly entering a period in which attacks on science led to attacks on individual scientists, singling us out as enemies of conservative viewpoints and beliefs. I thought about Henrik Ibsen's 1882 play, *An Enemy of the People* (*En folkefiende*), in which Dr. Stockmann, the village medical officer, is vilified for exposing the dangers of the local spa baths, which are contaminated with bacteria [54]. The parallels with my case seemed clear: I was a medical school professor in Texas yet was under attack by elected officials from other states [55].

February 1

February 1, 2022, will be a day that I remember for a long time. Even though Houston has a subtropical climate, it can still get cold in the winter, and I recall that ice and snow were in the forecast. But then some amazing news: That morning, Rep. Lizzie Fletcher, a member of Congress representing parts of Houston, announced that her office had nominated both me and my science partner for the past 20 years, Dr. Maria Elena Bottazzi, for the 2022 Nobel Peace Prize. This was an extraordinary and deeply meaningful honor, recognizing the efforts of our Texas Children's Hospital Center for Vaccine Development to make vaccines for poverty-related neglected diseases, most recently a low-cost recombinant protein vaccine for COVID-19. Moreover, we had developed this vaccine with no patents, even helping in the technology transfer (and co-development) of the vaccine technology to India, Indonesia, Bangladesh, and a new production facility starting up in Botswana. The Botswana initiative is a public-private partnership between the Botswana government (following our face-to-face meeting in Houston with President Mokgweetsi Masisi) and ImmunityBio, a US-based company headed by biotech entrepreneur Dr. Patrick Soon-Shiong (who grew up in South Africa).

I was elated at being nominated and was honestly thrilled by the recognition of the decades during which I had worked seven days a week, with the intensity ratcheting up even more during the pandemic. Getting our vaccine out to the countries that needed them placed an enormous work and emotional burden on our team and me personally. Now, with this announcement, I felt both relief and satisfaction.

Unfortunately, my relief was short-lived. Whether it was in response to my Nobel Peace Prize nomination or just a coincidence—I guess we will never know—for some reason that evening on Fox News, Tucker Carlson launched a very personal attack against me: "Hotez is a pediatrician who spent his life studying tropical parasites. He wouldn't seem to be the obvious go-to guest for cable news bookers looking for someone to speak knowledgeably about COVID.... Unfortunately for all of us, as Peter Hotez speaks, he discredits American medicine.... He's a misinformation machine constantly spewing insanity" [56].

Carlson conveniently forgot to mention that, yes, I conduct research on tropical parasites, but I also have extensive, hands-on expertise in coronaviruses and COVID-19 vaccines—indeed, likely far more experience than anyone who has appeared on Fox News. The following evening, a Stanford medical school professor who authored the *Wall Street Journal* article "Is the Coronavirus as Deadly as They Say?" appeared with Laura Ingraham on *The Ingraham Angle* to add, "Hotez is actually funded by Fauci's group. He's—much of his funding comes from the NIAID-NIH led by Fauci" [57]. What that doctor failed to mention is that NIAID-NIH is by far the largest funding source of infectious diseases research in the United States and globally. He also failed to disclose that the departments of microbiology and infectious diseases at his own institution, Stanford University School of Medicine, depend on NIAID-NIH support to conduct their research programs. I felt his statement was among the more disingenuous I had heard during the pandemic.

Overall, I was disgusted by these remarks. Not only were they either unfair or untrue, but also, when I first heard about them on social media the next morning, I knew what was coming. Predictably, death threats and celebrations of my impending execution followed these attacks. As

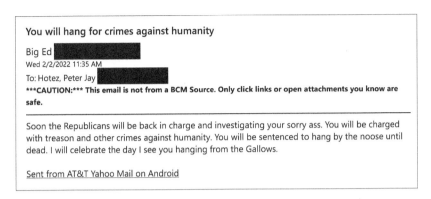

Figure 5.3. E-mail sent to me on February 2, 2022.

you can see from the time on the e-mail shown in figure 5.3, I did not have to wait long for the first ones to show up in my inbox.

The back-to-back Fox News programs signaled that I had joined Drs. Fauci and Daszak, along with other US biomedical scientists, as enemies of a political entity, namely, an authoritarian and far-right element of the Republican Party and conservatism in America. In my view, the members of Congress, governors, news outlets, and contrarian intellectuals committed to this movement are determined to halt or slow scientific progress, as well as intimidate the scientists. Their followers openly express their wish to see me die through a public execution—because I am a scientist.

We have entered a very dark and scary chapter in the intellectual life of the nation, one of a sort I thought had not been seen since the McCarthy era. During the 1950s, J. Robert Oppenheimer, the chief scientist of the Manhattan Project, fell under the scrutiny of the House Un-American Activities Committee (HUAC) before the Atomic Energy Commission revoked his security clearance and authority [58]. Oppenheimer was one of several left-leaning University of California at Berkeley faculty and students who had aroused HUAC attention because of their gatherings and political and philosophical discussions. The FBI began wiretapping their conversations, causing some to lose their teaching positions over the next decade. Just as the Cold War reshaped how American scientists interacted with the federal government [59], we may now be facing another such realignment. My concern is that the current anti-science political ecosystem and its links to major branches of the federal government represents state-sanctioned attacks on US scientists, just as the HUAC did 70 years ago.

Even though the current ecosystem of anti-science aggression has yet to employ full-fledged McCarthy era tactics, it remains a dangerous one for American science and scientists. I believe that the conservative news outlets, contrarian public intellectuals, House Freedom Caucus, and other elements in Congress, operating in tag team fashion, generate a synergy that fosters this menacing hostility [60]. In April 2021, Rep. Greene attempted to drive a portion of the House Freedom Caucus even further to the right through the establishment of a new "America First" caucus based on "common respect for uniquely Anglo-Saxon political traditions" [61]. Fortunately, this effort failed to gain the necessary support in the House of Representatives.

Nevertheless, by the end of 2021 and into 2022, these different elements began to reinforce one another in what appeared to be a well-organized and deliberate system. In addition to trying to undermine public confidence in biomedical science, these elements now also aspired to silence scientists like myself who publicly condemn anti-science aggression and explain how it was responsible for killing thousands of Americans during the pandemic. The methods and practices of these far-right attacks are not novel. Instead, they follow a playbook dating back to authoritarian regimes in the early twentieth century. Are we threatened by a homegrown authoritarian element that employs tactics used in the USSR when that country attacked scientists during and after Stalin's regime? The fact that this is happening in the United States in the twenty-first century is a chilling new reality.

Without science, democracy has no future.
—Maxim Gorky, April 1917

6 | The Authoritarian Playbook

The public attacks against me from high-profile elected officials or media personalities on the far-right often provoked a wave of hate e-mails and social media posts. Many also appeared after occasions when I was profiled negatively on Fox nighttime programming, typically *Tucker Carlson Tonight* and *The Ingraham Angle*, each ranked at the top of American television viewership. The type of rhetoric employed by Fox News provides the red meat its viewers crave. Shrill tones and free use of name-calling have become a Fox News signature.

Something else I have noticed about Fox News tactics: they have a familiar ring, at times even provoking flashbacks from my elementary school days when I first learned about Soviet propaganda. As a child growing up in New England in the 1960s, I still remember hearing stories about Nikita Khrushchev banging his shoe on the podium at the United Nations General Assembly or learning about the fiery exchanges between President John F. Kennedy and Khrushchev at their Vienna summit during the lead-up to the Cuban missile crisis. Cold War propaganda notoriously weaponized communication and liberally employed name-calling. Reminiscent of the Soviet era, the Fox News anchors called me "infamous" or referred to me as "a charlatan"—and were eager to quote me out of context. To me, such inflammatory statements resembled old-school Communist-style accusations.

Whether the similarities between the Fox News rhetoric and Soviet Russian propaganda were deliberate or coincidental is something we might never know. However, in the weeks following its broadcasters' attacks on me, it was curious how several news outlets and reporters strongly criticized Tucker Carlson for his support of Russian president Vladimir Putin and his generally pro-Russian views [1]. This became especially apparent as Russia mobilized troops in preparation for its invasion of Ukraine [2–4]. An April 2022 report in the *New York Times* found that references to Fox News stories and opinion pieces surged in the Russian media because they bolstered Russian conspiracy theories and criticized President Biden and the North Atlantic Treaty Organization [5]. Although Carlson defended himself against charges that he is a Russian propagandist [6], some prominent Americans in the media and other fields expressed outrage at his support of Putin, even to the point of asking the Department of Justice or other agencies of the federal government to investigate his pro-Russian activities [7, 8]. Something new and unsavory was happening as the US political right became increasingly enamored of authoritarian regimes in Russia or nations from the old Soviet bloc.

I also reflected on the awful legacy of Soviet Russian targeting of both physicians and scientists, remembering a heinous "Doctors' Plot" conspiracy of 1953. Shortly before his death, Stalin accused nine doctors, six of them Jewish, of deliberately attempting to harm Soviet leaders by administering dangerous treatments. Stalin himself allegedly edited an article that was published in *Pravda*, the official newspaper of the Soviet Communist Party, on January 13, 1953, titled "Ignoble Spies and Killers under the Mask of Professor-Doctors" [9]. A wave of hatred and anti-Semitism followed, with reports of Soviet citizens refusing medical care or treatments from Jewish doctors. Fortunately, the case against the doctors collapsed in the immediate aftermath of Stalin's death from a stroke.

One similarity between the old Soviet-era *Pravda* newspaper (or the TASS News Agency) and the tactics of the Fox News nighttime anchors was that each launched campaigns to discredit their physicians and scientists of national prominence. For both the Soviet press and Fox News, physicians and scientists represented threats to the state and therefore became bona fide targets of propaganda. Was I, together with colleagues, living through a modern version of the Jewish doctors' plot? Perhaps not exactly, but there are common threads, and the Fox News attacks were disconcerting in part because of the subsequent threats I received via e-mail, social media, or phone calls. They indicated some knowledge about where I worked, and I began to be concerned about my personal safety (and that of my family). In time, I sought advice and assistance from local law enforcement in Houston, and, because of the anti-Semitic content of these messages, from the Anti-Defamation League.

Science in the Totalitarian State

Linking anti-science attacks from the far-right to an earlier Soviet era has several historical implications. In his now classic 1941 *Foreign Affairs* article titled "Science in the Totalitarian State," the former *New York Times* science writer Waldemar Kaempffert examined the then-current state of scientific activities in Stalinist Russia and Nazi Germany [10]. He reported on the sharp decline in German science from its global supremacy in everything from physics to zoology during late nineteenth and early twentieth centuries to its collapse under Hitler. Central to the fall of German science was the exclusion of Jews from appointments at research universities because of what was termed "Jewish communism" or Jewish science. This Nazi discrimination included shunning Einstein's theories of relativity.

Stalin also repudiated relativity and cited it as an example of "bourgeois idealism." Indeed, his persecutions of Soviet physicists accelerated during the 1930s, a time known as the "great purge" of intellectuals considered enemies of the state. Between 1937 and 1938, more than 100 physicists in Leningrad (St. Petersburg) were arrested [9]. The arrests represented a piece of a larger effort by Stalin to extinguish intellectual life in Leningrad and to centralize power in Moscow. Ultimately, even the future Nobel laureate in physics, Lev Landau, fell under suspicion and was imprisoned, although he was eventually released. Less fortunate was Boris Mikhailovich Hessen, a Russian Jewish mathematician and ardent Marxist—but also champion of the theory of relativity—who died in a Soviet prison in 1938 [9]. Hessen impresses me as someone who tried to thread the needle, claiming that political leanings should have no bearing on science. Tragically, his claims that Einstein's theories of relativity had nothing to do with Marxism or Communism did not resonate with Soviet authorities, and he paid for them with his life.

I empathized with Hessen because, as he did, I worked hard to reach across the aisle and tried to depoliticize vaccines and other targets of anti-science aggression from the far-right. Well after I was excluded from Fox News primetime appearances in the evening, I did everything possible to accept invitations for daytime and weekend interviews. Moreover, I made myself available to NewsMax, another conservative news outlet, as well as to Breitbart, and I always accepted opportunities to meet with GOP elected officials. My major talking point: Everyone is entitled to conservative views, and in many cases even extremist viewpoints, but please do not embrace vaccine refusal or other aspects of health freedom propaganda, because it will only lead to losses of human life.

Soviet Genetics: The Legacies of Lysenko and Vavilov

Whereas both Stalinism and Nazism rejected relativity physics, Nazi science perversely accepted the concepts of Mendelian genetics—although in a corrupted form that led to heinous experiments on twins by Josef Mengele and others. As detailed below, Stalinist Russia completely scorned Mendelian genetics, a disdain that was directly responsible for the collapse of Soviet agriculture, and ultimately, the mass starvation and death of millions of Russians.

In *The Origins of Totalitarianism*, published in 1951, the philosopher and political scientist Hannah Arendt said this about intellectuals under totalitarian rule:

> Intellectual ... initiative is as dangerous to totalitarianism as the gangster initiative of the mob, and both are more dangerous than mere political opposition. The consistent persecution of every higher form of intellectual activity by the new mass leaders springs from more than their natural resentment against everything they cannot understand. Total domination does not allow for free initiative in any field of life, for any activity that is not entirely predictable. Totalitarianism in power invariably replaces all first-rate talents, regardless of their sympathies, with those crackpots and fools whose lack of intelligence and creativity is still the best guarantee of their loyalty. [11]

In the early and mid-twentieth century, totalitarian leaders and regimes waged war on science and scientists. Scientists were considered enemies, and the dictators found that maintaining power and totalitarian control depended on strong-arming scientists into submission. Such regimes also wound up replacing legitimate scientists with pseudoscientists or contrarian intellectuals subservient to the needs of the state. This may help us to understand why the Florida governor now seeks health policy advice from contrarians and so-called experts from far-right think tanks, many of whom make regular appearances on Fox News broadcasts.

There are many dark stories of great scientists whose lives were destroyed through totalitarian oppression, but few are more chilling than

Figure 6.1. Photo of the prisoner Nikolai Vavilov. Official photo from the file of the investigation. *Source*: The People's Commissariat for Internal Affairs, Central Archive of the Federal Security Service of the Russian Federation (Moscow), Institute of Plant Industry, created January 1, 1942, https://doi.org/10.1371/journal.pbio.3001068.g001.

the plight of Nikolai Vavilov, one of the leading Russian geneticists of his generation (fig. 6.1) [9, 12, 13]. In the early twentieth century, Vavilov was a scientific superstar with an international reputation. He pioneered Mendelian genetic approaches to agricultural botany, leading to improvements in cereal crop yields. His research to develop and breed seeds that could withstand freezing temperatures was essential for producing hardy crops during harsh Russian winters. Vavilov, as one of the Soviet Union's most distinguished scientists, served as the director of the Lenin All Union Academy of Agricultural Sciences. The academy led national efforts to improve crop yields, also serving as the scientific wing of the Soviet Commissariat of Agriculture [9].

Despite, or perhaps because of, Vavilov's scientific stature and accomplishments, he was eventually undermined by Trofim Lysenko, a man with no doctoral degree or rigorous formal training, yet someone who gained favor with Stalin for refuting the bourgeois idealism of Mendelian genetics in favor of pseudoscience. Lysenko had put forward a concept known as vernalization, considered a form of long-discredited Lamarckian theories. Jean-Baptiste Lamarck was an eighteenth-century

French zoologist who hypothesized that the inheritance of physical features, such as an organ or limb, depended on its physiologic use during the life of the organism. Conversely, the failure of inheritance was a consequence of disuse or attrition. Through the vernalization process, Lysenko proposed soaking the seeds of winter wheat in water and later burying them in the snow. This would keep the wheat accustomed to the cold and make it hardy prior to its final planting in the spring. As Simon Ings quipped in *Stalin and the Scientists*, vernalization would make it possible to "grow alligator pears and bananas in New York and lemons in New England" [9]. The facts that Lysenko was a Russian peasant without a formal scientific education and that he rejected Mendelian genetics made his activities attractive to Stalin and the Soviet Communist Party during the 1930s.

As Lysenko ascended in the Soviet scientific hierarchy, he convinced the Soviet leadership to abandon Mendelian genetics in favor Lamarckian vernalization theories. In 1936, the Soviet leadership denounced Mendelian genetics as "fascist," although this may have been engendered in part by some influential scientists believing Mendelian theories might lead to eugenics, as promoted by Nazism. Nonetheless, an aggressive purge of biological scientists with alleged fascist links ensued. The American *Drosophila* geneticist (and future Nobel laureate), Hermann Muller, who had left the University of Texas and headed to Europe before receiving an invitation to organize genetics research in the Soviet Union in 1933, left abruptly in 1937. That same year, Solomon Levit was forced out as the director of the Medical-Biological Institute [9]. He was later arrested and shot. Other genetic researchers suffered a similar fate. By 1939, Vavilov and Lysenko were in open opposition and had become bitter rivals. According to Vavilov's biographer, Peter Pringle, Vavilov would remark that Lysenko's rejection of Mendelian genetics ran "counter to all of modern biology" [12]. Vavilov worried that Lysenko's actions would transform the Soviet Union into a backwater of science and isolate it from the global scientific community. Tensions between the two mounted, and during an agricultural fair Vavilov confronted Lysenko, cursing him for destroying Soviet science. However, Stalin clearly backed Lysenko, and he summoned Vavilov to his Kremlin office at 10 p.m. on November 30, 1939. Sitting at his desk Stalin was alleged to have said, "Well citizen Vavilov, how long are you going to go on fooling with flowers and other nonsense? When will you start raising crop yields?" [12].

Although by 1940 the great purge was winding down, Vavilov knew he had become a potential Kremlin target. The NKVD (the People's Commissariat for Internal Affairs and also the secret police) arrested Vavilov while he was on a plant-hunting expedition in Ukraine and brought him to Moscow for imprisonment. By then, Lysenko had replaced Vavilov as the president of the Lenin Academy of Agricultural Sciences. A court sentenced Vavilov to death, a ruling that was later commuted after Vavilov wrote personally to Lavrentiy Beria, Stalin's infamous Soviet deputy premier and chief of security, asking him for clemency. However, Vavilov remained in prison, and in 1941, ahead of an advancing German army, he was loaded on a prisoner train to Saratov. By 1943, Vavilov was suffering extreme malnutrition in prison and, despite multiple appeals for his release from the international community and even British prime minister Winston Churchill, he died in a prison hospital. As Pringle appropriately lamented, a scientist who devoted his life to improving crop yields was starved to death, only to allow his successor to advance a program of pseudoscience that caused crop failures leading to the deaths of an untold number of Soviet Russian peasants [12].

This time in Communist Russia represents the fullest modern expression of anti-science, but we can see how its elements linger in the present. During the COVID-19 pandemic, Fox News amplified the views of the contrarian intellectuals or elected officials who downplayed the importance of masks or sought to discredit COVID-19 vaccinations by questioning their effectiveness and safety. Thousands of Americans accepted their propaganda and perished.

Following Stalin's death, Nikita Khrushchev pardoned Vavilov posthumously, but the persecution of scientists did not end. Most notably, the renowned physicist and architect of the Soviet hydrogen bomb and later peace and human rights activist, Dr. Andrei Sakharov, underwent arrest, forced exile, and house detention in Gorki, a city forbidden to foreigners at the time [14]. Earlier, Sakharov had won the 1975 Nobel Peace Prize, and on the day it was announced, Soviet leader Yuri Andropov claimed that the prize was merely a political instrument intended to discredit the USSR. He claimed it was awarded "for provocative purposes, in order to support his anti-Soviet activities and on that basis to consolidate hostile-minded elements within the country" [14]. The Kremlin refused Sakharov permission to leave the country to attend the Nobel ceremony in Oslo. Yelena Bonner, his wife, instead delivered the acceptance address, titled "Peace, Progress, Human Rights," which called for an end to the nuclear arms race between nations, while promoting an agenda of human rights. During a follow-on KGB assembly meeting, Andropov labeled Sakharov "Public Enemy No. 1." Sakharov's health suffered, and he died from heart failure in 1989 at the age of 68. Today, an asteroid that circles the sun in a belt between Mars and Jupiter is named in his honor.

Biomedical Science and Modern Authoritarian Regimes

When I receive threats from those who identify as "patriots," I remind myself that American scientific research helped the nation defeat fascism in World War II, achieve victory in the Cold War, and begin efforts to defeat catastrophic illnesses such as cancer and HIV/AIDS. The scientists are the true American patriots. As Kaempffert correctly pointed out in the months before the United States entered World War II, "If the world wants to preserve science as a powerful social force for good, the research physicist, chemist and biologist must be permitted to work without intellectual restraint, i.e. to enjoy the fundamental freedom of democracy" [10]. In modern times, the US government has maintained a strong track record of supporting US scientists through the NIH and

National Science Foundation, the establishment of a White House Office of Science and Technology Policy, and its recognition of the National Academies of Science, Engineering, and Medicine—but there have been some notable missteps. These included the attacks on J. Robert Oppenheimer and other atomic scientists during the McCarthy era and the withering attacks on climate scientists.

Today, it is a rarity for world leaders to sentence prominent scientists to perish in gulags or execute them for their academic or intellectual pursuits, although this practice continues in Iran, North Korea, and other extremist regimes [15]. However, many scientists remain under severe threats or persecution by autocrats or authoritarian leaders across the globe. In her post–World War II treatise, Hannah Arendt distinguished between totalitarian and autocratic or authoritarian regimes [11]. Autocracies and authoritarian regimes marginalize, suppress, or even make illegal political opposition, but they still allow certain political, social, or educational institutions some uneasy degree of coexistence. In contrast, totalitarians typically seek to control almost every aspect of people's lives and are willing to use summary executions to terrorize populations into submission. Both totalitarian and authoritarian regimes may view some of their most prominent scientists as enemies of the state, along with a larger group of intellectuals in the humanities and arts. Arendt has also noted how the best scientists and intellectuals under autocrats and totalitarian dictators were replaced at national academies or universities with second-rate talents who proved their worth only through their state loyalty.

The journalist and historian Anne Applebaum observes how science and scientists in eastern Europe can become targets of authoritarian regimes in modern times [16]. In recent years, Hungary has fallen under increasingly autocratic rule by its authoritarian ruler, Viktor Orbán. Applebaum reports that the Orbán regime does not simply dissolve the Hungarian Academy of Sciences but instead replaces its qualified scientists and leaders with those hand-chosen to show loyalty to the regime. Presumably, neutralizing scientists and other intellectuals is recognized as central to the survival of autocratic governments, much as was believed by the totalitarian regimes of the past. Orbán also shut down the Central European University, funded by George Soros, and forced it into exile in Vienna [17]. At the time this happened, the Central European University ranked among the top universities in Europe and was probably the best in central Europe. Orbán dissolved a national treasure and sent it into exile because it represented a potential political threat.

Now are we are beginning to see a confluence of Hungary's authoritarian aspirations with those who extol far-right and autocratic practices in the United States. It was especially chilling to watch Tucker Carlson travel to Hungary during the summer of 2021 to applaud the virtues of the Orbán regime, later releasing a new documentary film titled *Hungary vs. Soros: Fight for Civilization* [18]. Carlson's pivot to Orbán's controlling style of government reinforced my impressions that his February 2022 attack against me was born out of authoritarian leanings. These activities also preceded a spring 2022 CPAC conference in Budapest, which included a recorded message from Carlson and another speaker calling for "American Orbanism" [19], followed by CPAC's invitation to Orbán to speak at its second 2022 summit in Dallas, co-attended by prominent anti-vaccine activists. According to one report about Orbán's remarks, "He repeatedly pandered to the Texan audience, calling Hungary 'the lone star state of Europe'" [20]. Princeton University's Kim Lane Scheppele, who studies the politics of Hungary, observes, "Hungary has become, for the Trumpist Republicans, what Sweden used to be for the social democrats—it's proof of concept" [21]. In this sense, a new global alliance might unfold between political extremists and conservatives in the United States and the autocrats in Hungary. Among other things, they are bound by a common grievance against science and scientists.

It is also noteworthy that CPAC Brasil held its second conference in 2021 in Brasilia, the capital, and the first in São Paulo in 2019 [22]. Both summits reflect the pursuits common to the far-right in the United States and the authoritarian regime led by Brazilian then-president Jair

Bolsonaro, who made major cuts in funding support for Brazil's major research universities and institutes after taking office in 2019 [23]. Under Bolsonaro, Brazil's leadership has been reported to make heavy use of Telegram, a social media platform favored by far-right extremists because of its use of encryption and minimal oversights and regulation [24, 25]. This and other vehicles promote attacks against the opposition party and discredit vaccines. Throughout the pandemic, Brazil's scientific establishment heavily criticized the Bolsonaro regime for downplaying the severity of the pandemic, refusing to implement COVID-19 prevention measures, promoting unproven treatments, and delaying the rollout of vaccinations. Just as Americans lost their lives because of anti-science aggression, some estimates indicate one half of the lives lost in Brazil's COVID-19 epidemic were preventable [23]. The enforcement of authoritarian or autocratic rule is tightly aligned with efforts to crush science and scientists, and this became especially obvious for the biomedical sciences during the pandemic. We should not expect this situation to abate even after the pandemic ends.

In a podcast with New York University history professor Ruth Ben-Ghiat [26], an expert on authoritarianism in the United States and the author of *Strongmen: Mussolini to the Present* [27], and later with Molly Jong-Fast, a journalist and commentator who previously hosted her *New Abnormal* podcast with the *Daily Beast* [28], I tried to elucidate this new authoritarian assault on prominent US biomedical scientists. My worry is that the attacks against us are not random, and they certainly do not arise from grassroots, "mom and pop" groups. Rather, the assault on American science and scientists from the triumvirate—conservative news outlets, far-right members of the US Senate and House of Representatives, and contrarian intellectuals or pseudointellectuals—comprise elements of an emerging far-right authoritarian movement that is every bit as aggressive and dangerous as what Orbán engineered in Hungary or Bolsonaro in Brazil. The transcript from my podcast interview with Ruth Ben-Ghiat on this issue is instructive:

BEN-GHIAT: This is what I call the upside-down world of authoritarianism, when those who are trying to save our lives become the ones you're supposed to eliminate.

HOTEZ: I don't even call it anti-vaccine misinformation or disinformation anymore. I call it anti-science aggression. It's tough for someone like me to talk about it because our training as scientists says that we're supposed to keep our heads down and focus on the science. But there's nobody speaking out. We're not hearing from the professional societies, the academic societies, the national academies about this issue. These are not mom and pop groups we're up against. These are well-funded, well-organized political entities that are working to undermine the fundamental infrastructure of American science. And we've got to take a stand.

BEN-GHIAT: When you study the rise of fascism, in fact, you see how professional, scientific, and historical academies all tried to remain "objective" until it was too late.

HOTEZ: I'm trying to get my colleagues to see that this goes beyond the expertise of the health sector. A hundred thousand Americans lost their lives over the summer from the Delta variant, despite the widespread availability of vaccines. These are people who have chosen to take their own lives and refuse vaccinations. So this is a killer. [26]

Our exchange highlights how the anti-vaccine movement evolved from one focused on health disinformation to become an essential element of extremist politics. Extremists within the GOP adopted anti-vaccine sentiments and anti-science as a major platform. In time, expressing anti-vaccine and anti-science viewpoints became a centerpiece of the far-right elements of the GOP, whose authoritarian expressions play out daily. A similar situation is under way among authoritarian regimes in Hungary and Brazil, but it is not limited to these two nations. For example, to varying degrees autocratic leaders in the Philippines, El Salvador, and Nicaragua have adopted anti-science platforms, which one writer terms "medical populism" [29]. In each case, such regimes downplayed the severity of the COVID-19 pandemic, established a low bar for the level of past infection or immunizations required for herd immunity, and hyped miracle cures such as ivermectin or hydroxychloroquine. They also discouraged mainstream COVID-19 prevention measures related to reporting, contact tracing, masks, and ultimately vaccinations. Each of their actions included primitive efforts to divert attention away from the COVID-19 pandemic or their failed responses to it. Far-right candidates in the 2022 French presidential election also openly opposed vaccine mandates as they challenged their very pro-vaccine incumbent, President Emmanuel Macron [30]. This situation creates a new expectation, as conservative candidates see opposition to science and scientists as central to their platforms. From its health freedom propaganda beginnings almost a decade ago in Texas and elsewhere in the United States, the anti-science empire now pervades multiple authoritarian regimes in Latin America, Europe, and Asia, espousing a set of principles centered on discrediting science and scientists.

A Final Twist: Russia Weaponizes Health Communication

Layering on top of the complexity of an authoritarian embrace of anti-science are the emerging findings of how Russia created a program of anti-vaccine disinformation as a weapon to destabilize democracies. In 2021, reports from the UK Foreign Office and US State Department highlighted Russian involvement in discrediting major COVID-19 vaccines produced by Western multinational companies [31]. In some cases, according to sources, Russia did this to promote its own homegrown adenovirus-vectored vaccine produced by the Gamaleya Research Institute and known as Sputnik V [32]. Russian disinformation also went beyond COVID-19 vaccinations during the pandemic and led to efforts to spread false rumors that the virus was created deliberately as a bioweapon or that it was the product of 5G radio waves [33]. Its propaganda campaigns employ multiple channels and media approaches in a blitz that is sometimes referred to as a "firehose of falsehood." The messaging is described as high volume, multichannel, repetitive, and without consistency or even reality [34]. Additional reports find that Russia was not alone. China was a major instigator of COVID-19 disinformation during the pandemic, according to the European Union [35], while the US government alleges the same for Iran [36].

That Russian propaganda was working to undermine North American and European COVID-19 vaccines or other prevention efforts should come as no surprise. Even before the pandemic, George Washington University's Dr. David Broniatowski reported in 2018 on an elaborate system of computer-generated bots and trolls that Russia used to sow doubt about childhood vaccinations. He found that Russia uses anti-vaccine computer algorithms to generate disinformation and disseminate it in ways to create internal conflicts and amplify disagreements between pro-vaccine and anti-vaccine groups in the United States and other democratic regimes [37]. Along similar lines, the *New York Times* reported in 2020 (later updated in 2021) on US State Department findings that Russia uses social media accounts numbering in the thousands to promote COVID-19 disinformation [38]. Russian health disinformation dates to the beginning of the Putin regime in the early 2000s. According to the *New York Times*, back then, Russia attempted to use social media and other platforms, including its US propaganda arm, previously known as Russia Today and now simply as "RT," to claim that the United States bioengineered the HIV/AIDS virus, H1N1 pandemic flu, and the Ebola virus as agents of biowarfare [38]. In other cases, Russia attempted to create false rumors that the United States was experimenting on African populations through weaponization of either the actual pathogens or the vaccines developed against them. Broniatowski shows how in some cases Russian propaganda uses both pro-vaccine and anti-vaccine messages when its goals are more about promoting discord and confusion rather than bona fide interest in the vaccine issue. In some instances, this tactic has proven to be a successful propaganda technique for exploiting hot-button issues to promote instability. In another odd twist, Kiera Butler at *Mother Jones* has also reported that prominent wellness influencers in the United States now help disseminate both anti-vaccine and Russian propaganda, including disinformation about Ukraine as a justification for the Russian invasion [39]. It seems that disinformation comes in some sort of dystopian package that combines anti-vaccine activism, pedaling products for health and wellness, and support for autocratic governments.

As a final irony, perhaps no one has suffered more from Russia's disinformation empire than Russia itself, as COVID-19 vaccine refusal and hesitancy rates in Russia rank among the worst [40]. Many experts agree that the official death toll from COVID-19 in Russia likely reflects either gross underreporting or deliberate misrepresentation by the Putin government [41]. Officially, Russia has suffered the fourth-largest number of deaths after the United States, India, and Brazil. However, the excess deaths in Russia greatly exceed those reported. An analysis by the UW-IHME finds that as many as 1.07 million Russians died in the COVID-19 pandemic in the years 2020 and 2021 [42], or roughly the same number of deaths that occurred in the United States, but in a nation with less than one-half the population. This places Russia as the worst-affected nation in terms of excess COVID-19 mortality rates.

Halting Authoritarian Anti-science

Finding strategies to slow the spread of anti-science by authoritarian regimes or entities has become one of our great challenges. While the US government and Office of the Surgeon General focus on Facebook or the other social media companies in spreading misinformation, few governments or United Nations agencies wish to confront the source.

Therefore, anti-vaccine or health disinformation generated by Russia and other authoritarian governments now proceeds without significant interference. In a previous book and other writings, I presented opportunities to rebuild joint US-Russian international collaborations around vaccines. During the Cold War, programs in vaccine diplomacy between America and the USSR gave us new vaccines for polio and improvements that led to the eradication of smallpox [43]. However, reviving these collaborations does not appear likely in this time of Russian aggression toward Ukraine and other republics that were created after the collapse of the Soviet Union. As the State Department and major US intelligence agencies work to diffuse Russian bots and trolls, there is still no national plan to confront anti-science aggression from the far-right and authoritarian regimes. We now face our own internal authoritarian ecosystem whose leaders portray scientists as threats. Some political scientists express concerns that such activities, especially in the context of the January 6, 2021, storming of the US Capitol, threaten the future of democracy in the country. One writer refers to the United States as an "anocracy," caught between weakened democratic institutions and far-right authoritarianism [44]. These authoritarian trends now linked to anti-science have globalized to Brazil and Hungary, and threaten other nations as well. Anti-science could become a dominant feature of a new world order. This is not easily undone.

We must take sides. Neutrality helps the oppressor, never the victim. Silence encourages the tormentor, never the tormented. Sometimes we must interfere. When human lives are endangered . . . sensitivities become irrelevant.
—Elie Wiesel, *The Night Trilogy: Night, Dawn, The Accident*

MAGIC THEATER. ENTRANCE NOT FOR EVERYBODY.
—Hermann Hesse, *Der Steppenwolf*

7 | The Hardest Science Communication Ever

As terrible as it is that 200,000 unvaccinated Americans lost their lives needlessly despite widely available COVID-19 vaccines, we must anticipate that the legacy of anti-science aggression will continue beyond that haunting statistic. In some respects, the horrific loss of life might represent just the start of a cascade of adverse health and socioeconomic declines. Many of those who lost their lives because they refused a COVID-19 immunization were parents, grandparents, or other caregivers. An April 2022 article by the science writer Tim Requarth points out that on average, many American public schools now have at least one or two children who lost a parent or caregiver in the pandemic [1]. In addition, the loss of a caregiver during the COVID-19 pandemic now accounts for 1 in 12 orphaned children in the United States, and we already know that orphans face increased odds of poverty, dropping out of school, substance abuse, and suicide risk. In this way, COVID-19 will produce broad-reaching socioeconomic consequences for individuals, families, communities, and nations.

Then there are the public health consequences of long COVID (also referred to as post-acute COVID-19 syndrome) possibly lasting months, or even years, which is now an important cause of disability in the United States. The higher rates of severe COVID-19 cases among the unvaccinated living in conservative areas of the American South, Appalachia, and elsewhere may also mean greater numbers of long COVID sufferers. Our nation's already depleted health system now faces the prospect of managing the long-term care of millions of Americans afflicted by long COVID cardiopulmonary and renal insufficiency and neurocognitive disturbances [2].

Especially worrisome are the findings from Oxford University researchers showing gray matter brain degeneration from long COVID, with associated cognitive impairments [3]. Such neurologic damage across large segments of the US population might also have been prevented if vaccines were accepted. There is an increasing body of evidence to suggest that COVID-19 vaccinations not only keep individuals out of hospitals and ICUs and prevent deaths but also reduce the frequency and impact of long COVID [4, 5]. The bottom line: We have not even begun to imagine the scope and scale of the mental health devastation that will result from long COVID, loss of parents and other caregivers, and heightened levels of anxiety from a traumatized American public [6]. This occurred in no small measure because a critical mass of Americans refused COVID-19 vaccinations.

Long-Lasting Damage to American Biomedical Science

We must also be mindful that an empowered and well-organized anti-vaccine movement in America will not disappear after our COVID-19 pandemic ends. Through their abilities to organize, secure funds, dominate the Internet, and make themselves essential to far-right authoritarianism, anti-vaccine activists now hold enormous power in the United States. The anti-vaccine lobby will continue to accelerate this trajectory.

To begin, they will expand their efforts to reduce child and adult vaccination rates for non-COVID conditions. An Economist/YouGov poll conducted in fall 2021 found that traditional support for pediatric vaccinations required for school entry and attendance has waned considerably during the pandemic, with the decline occurring along a red/blue partisan divide, similar to COVID-19 vaccine resistance [7]. Only 46% of Republicans now favor required vaccinations for school-aged children, down from 59% the year before, and far less than the 85% of Democrats polled who support school-required vaccinations. According to Dr. John Moore, a microbiologist at Weill Cornell Medical College in New York, "Since the pandemic began, we're also seeing more politics-driven attacks on state mandates for pre-school vaccination. Long-vanquished child killing diseases will rise again, just because parents have been fooled into rejecting safe, long-proven vaccines" [8]. Because most vaccine policies for school entry or attendance are set by state legislatures rather than the federal government, we must anticipate a new normal in which pediatric vaccine mandates are peeled back in the conservative red states.

In 2021, Tennessee Republican legislators led efforts to pressure their state health commissioner to fire Dr. Shelley Fiscus as the vaccination director because of her support for pediatric COVID-19 vaccinations [9]. It is likely that other state vaccine directors or health officials will face similar pressures to renounce existing or additional school-entry vaccine requirements. Assuming this trend continues, we might see the public health consequences in quick order. Early data are already beginning to show declines in routine child immunization rates in the conservative red states [10]. Simply put, we should expect the return of some childhood infections that had previously been eliminated from the United States. The first is likely to be measles, owing to its high transmissibility and the fact that measles reemergence was already beginning in the year prior to the pandemic because of accelerated anti-vaccine activities [11].

Given the spread of health freedom sentiments to Canada and western Europe there could be a domino effect. For instance, the *BMJ* (*British Medical Journal*) has already noted a decline in parental acceptance of routine childhood vaccinations in the United Kingdom since the start of the pandemic [12]. In the meantime, the World Health Organization (WHO) has sounded an alarm regarding global vaccination coverage, noting significant declines since the start of the pandemic [13]. Undoubtedly, some of its findings, including a 5% drop in general immunization coverage and an increase in the number of completely unvaccinated children by 5 million, reflect the social disruptions caused by the pandemic. However, even with catch-up vaccination campaigns after the pandemic subsides, we might not return to the baseline because of heightened anti-vaccine sentiments. In this sense, an acceleration of anti-vaccine activities could slow, halt, or even reverse many of the gains made over the past year that were achieved through Gavi, the Vaccine Alliance, and the United Nations' Millennium Development Goals and Sustainable Development Goals [14]. If immunization levels decline, we might see the widespread return of measles, or even polio.

Both the WHO and UNICEF report that globally measles cases rose by almost 80% in the first two months of 2022, compared with the first two months of the prior year [15]. The countries with the largest measles outbreaks include the conflict or post-conflict countries of Afghanistan, Somalia, and Yemen, as well as Ivory Coast and Liberia in Africa [15]. This follows a pattern of measles epidemics in countries with fragile health systems and immunization gaps, but if anti-vaccine sentiments and activities now also pervade such nations, this could lead to an explosion in measles cases.

Documenting the impact of anti-vaccine activism from US conservatives on low-income African nations is not easy, especially because much of it circulates through local social media, newspapers, or even word of mouth. However, several footprints of American anti-vaccine activities on the African continent have been discovered, including the use of similar rhetoric, talking points, and even memes, as well as references to Fox News and the remarks made by US political leaders from the far-right [14]. In central Asia, Afghanistan and Pakistan currently rank among the most polio vaccine–hesitant areas of the world. Of particular concern in these two countries are the tragic attempts in 2021–22, some of them successful, to assassinate polio vaccinators [16]. Although the forces driving this level of vaccine extremism differ from those in North America and Europe, I am concerned about the potential confluence of anti-vaccine aggression between the Eastern and Western Hemispheres. This would contribute to a profound undermining of efforts to eliminate or eradicate this crippling disease.

When it comes to controlling global measles and polio, the return of these illnesses is not restricted to poor countries. As noted above, measles almost returned to the United States in 2019, as it had done earlier in Europe [11]. Then in 2022, the United States saw its first polio case since 1993, while in the United States and the United Kingdom, poliovirus was isolated from urban wastewater [17–21]. These instances may represent more than one-off findings and might reflect further declines in vaccination. Poliovirus strains continue to circulate in the environment primarily because gaps in vaccination facilitate ongoing transmission. In the US and UK cases, the poliovirus discovered was derived from a strain that originated from the live oral vaccine (vaccine-derived poliovirus, or VDPV) but mutated until it acquired characteristics that resembled a wild-type poliovirus. It can then propagate among the unvaccinated. Therefore, the presence of VDPV is a biomarker for significant numbers of unvaccinated people. In 2022, an unvaccinated man from New York actually contracted paralytic polio [19–21]. These findings should be interpreted as a warning of possible additional cases unless efforts are implemented to redouble immunizations. The Global Polio Eradication Initiative has been one of the most successful international cooperative efforts toward disease elimination, and one with the potential to wipe out an ancient disease in the same manner that successful global smallpox eradication programs did during the 1960s and '70s. Failure to complete the eradication of polio because of anti-vaccine activism would be distressing at multiple levels, but especially because the victory it would signal for the anti-vaxxers is merely a pyrrhic one.

Still another concern is the potential for monkeypox to gain a foothold in the United States. In 2022, a large epidemic emerged, and there is a danger that it might generalize across the American population and even become enzootic, meaning that it circulates among rodent or other animal populations. The best way to halt the further spread of monkeypox is through timely and expanded immunization with a vaccine that the US Food and Drug Administration approved before the pandemic. However, there is potential for US anti-vaccine activists to target this vaccine and discourage its use [22].

Authoritarian trends emanating from the far-right will also extend beyond the vaccine space or other infectious disease prevention efforts. We should expect it to infiltrate and undermine other aspects of the biomedical sciences. This process began prior to the pandemic. Starting in 2017, the Boston-based Union of Concerned Scientists identified at least 100 instances in which the Trump administration attacked science within the federal government, including the targeting of individual government scientists [23]. One of the more detrimental elements has been the ban on fetal tissue research, which is essential for learning how Zika virus infection affects the fetal brain, or for fundamental research on transplantation, stem cells, HIV/AIDS, Parkinson's disease, and even COVID-19 [24]. Potentially, the bans on such research might eventually halt progress in additional cutting-edge technologies such as gene editing, systems biology, or the latest in computational biology or bioinformatics [25]. Programs of public engagement on a full range of these life science technologies and their applications will be essential to prevent these important areas from falling victim to the anti-science forces that plague the spheres of vaccines and vaccinations [26]. Otherwise, the United States risks falling behind other advanced nations in science.

Communicating Science

As health freedom propaganda accelerated in the United States during the previous decade, it became clear that the counteroffensive to halt its progress was insufficient. Private nonprofit and government-led vaccine advocacy groups made heroic efforts to promote positive vaccine messages and provide timely and accurate vaccine information to the public. However, such pro-vaccine advocacy needed parallel efforts to confront anti-vaccine and anti-science aggression and its political ties to conservative politicians, news outlets, and other far-right elements. Health freedom politics proceeded mostly without strong opposition. Then there was the community of professional scientists. While the biomedical scientific community was not exactly invisible, it often lacked the drive and capacity to work aggressively and strategically to dismantle anti-vaccine and anti-science activities.

Arguably, this has been done before but not sustained. The March for Science was held on Earth Day, April 22, 2017, in Washington, DC, and other cities globally. Smaller rallies and marches followed in 2018 and 2019. While a number of scientific societies, including the American Association for the Advancement of Science (AAAS), held side events and parallel activities warning about the rise of anti-science attitudes, our profession overall could have and should have responded sooner and with more visibility, volume, and heft, especially in response to anti-vaccine activism. In the biomedical sciences in America, we paid a steep price in the form of a public that too often disregards the science or questions the integrity, motivation, and sincerity of our professional activities. We even endured targeted threats and attacks.

Therefore, scientific communication and public engagement represent key areas for expansion and improvement. Even before we knew the pandemic was about to unfold in the United States, I sounded an alarm that scientists need to communicate better and become more visible in the public eye [25]. Only about one in four Americans could actually name a living a scientist, according to a study commissioned by Research!America, a Washington, DC–based policy and advocacy group. Moreover, among those who could come up with a name, it was typically someone like Neil deGrasse Tyson or Bill Nye, who are both outstanding science communicators [27], but otherwise our scientific community had become invisible to the public. The average American has no idea of what we do on a daily basis. This could breed misunderstanding or even mistrust and undermine our profession.

The American Academy of Arts and Sciences through its Public Face of Science Initiative, as well as the AAAS and the National Academies of Science, Engineering, and Medicine recognized the problem and hosted workshops and other events to encourage science communication by its practitioners. Such efforts were in direct response to the Trump administration's attacks on government scientists.

I proposed programs to introduce science communication training in doctoral and postdoctoral science programs, recognizing that scientists must be taught these skills if they are to be successful in reaching out to the public [25]. This includes employing modern approaches to communication that go beyond simply providing facts to counter misinformation. As Dr. Valerie Reyna, a Cornell psychologist who investigates effective science communication, points out, we urgently need to communicate with emotion and appeals to core human values rather than just present facts [28]. Similarly, Dr. Timothy Caulfield from the University of Alberta emphasizes targeting personal identities and ideologies [29]. These recent findings could become critically important in combating anti-science. Dr. Dietram Scheufele and his colleagues who specialize in life science communication at the University of Wisconsin highlight how we need to apply evidence-based approaches to public engagement in science communication, with an emphasis on partnerships between scientists and STEM practitioners, examination of theory-based and "practice-informed" hypothesis-testing, hard metrics, and clearly defined goals [26, 30]. Storytelling can also be useful and powerful. When I explain why my daughter Rachel's autism was not caused by vaccinations but through the actions of a specific gene involved in neuronal communication or trafficking [31], or when Dr. Kay Jamison, a Johns Hopkins University clinical psychologist, explores bipolar disorder through her own unique lens [32], the information often resonates especially well with the public; it also portrays us as real and relatable people.

However, although training scientists to communicate better and more frequently is essential, it is not sufficient to compete with an expanding anti-science universe. Success in this endeavor also requires a fundamental shift in our academic culture, which too often still clings to twentieth-century views that public communication by scientists represents some form of self-aggrandizement. As the world of communications and interconnectedness undergoes wrenching changes, the old-fashioned idea that scientists should keep their heads down and stick to their experiments, published papers, and grants no longer works. Such guidelines for our scientific profession were put in place before the arrival of the Internet and social media. These earlier attitudes created generations of silent scientists and resulted in a terrible vacuum. Anti-science now fills the gap. It was not a surprise to me, therefore, that when scientists and medical professionals did begin speaking out to educate the public during the COVID-19 pandemic, this overture was met by a backlash from anti-science groups and even segments of the public. Two surveys conducted during the pandemic by *Science* and *Nature* magazines, respectively, found that many researchers who published regularly about COVID-19 experienced harassment online and through other mechanisms [33, 34]. In one of the studies, 70% reported a negative impact, including physical threats, following a media interview or if they posted on social media [34].

What might a culture shift that encourages science communication look like? In addition to offering science communication training in doctoral and postdoctoral training programs, we might consider incentivizing young faculty members in our research universities and academic health centers to speak out and engage. I am evaluated annually, just like any medical school faculty member, but I am mostly asked about my extramural grant funding and scientific papers, especially those that are published in journals with "high impact." This term refers to a metric reported by the Clarivate Web of Science group and other organizations to rank the quality of scientific journals. Although controversial, the reality is that many academic institutions evaluate their scientists on this basis. In contrast, I am seldom asked to summarize my public engagement interests and activities. My experience is not unique. At many US institutions, there is very little interest in encouraging trainee or faculty public engagement activities, such as writing opinion and commentary articles, cable television and radio appearances, and certainly not posting on social media. While the institutions in our Texas Medical Center are better than most, the message too often sent by leaders of research universities and academic health centers is that public engagement is tolerated, but barely, and certainly not welcomed. Such attitudes have a particularly chilling effect on young scientists, who often receive the message that they engage the public at their own risk. Most university and academic health center officials would probably think of this as risk mitigation to protect the institution. Too often, it becomes a sort of a modern day "sword of Damocles" hanging over our heads—an inner voice that says "don't screw this up and possibly bring negative attention to the institution"—that mostly silences biomedical scientists.

We need to flip this issue on its side and think about how to shape evaluation metrics that actively encourage and do not discourage science outreach and public engagement. Public engagement and science communication should be prioritized as a vital and essential activity for biomedical scientists. Certainly, not all scientists want to enter the public arena—and that is fine as well—but for those wishing to engage, such an option should exist, or systems might even be established to support it. In my view, young scientists' commitment to public service and engagement is at an all-time high, and we should actively seek to capture and encourage this enthusiasm.

Science and Human Values

It might not be intuitive why, as an MD PhD laboratory investigator who develops vaccines for global health, I should also take on the challenge of combating anti-science or authoritarian regimes. Some colleagues, especially those who have been scientists for decades, even disapprove of my taking a public stance. Their position is a traditional one, that countering anti-vaccine movements or defending science and scientists is best left to others, while speaking out to defend the discipline is even inappropriate for our profession.

But I believe that confronting anti-science movements is essential and necessary [35]. Our training as biomedical scientists gives us the knowledge and tools to advance the fields of biochemistry and molecular biology, microbiology, neuroscience, pharmacology, and physiology. Many of us even apply our training to develop novel and exciting technologies, such as new medicines, diagnostics, medical devices, or vaccines. A common denominator is that we use science to save lives. I embarked on MD PhD training to apply the then-new science of molecular biology—the first gene had been cloned just a few years earlier—to the study of medically important parasites in Africa and elsewhere [36]. I brought hookworms into the Rockefeller University Laboratory of Medical Biochemistry (headed by Anthony Cerami), to take a first stab at developing hookworm vaccines. Now we have a human hookworm vaccine in clinical trials in Africa [37], as well as a vaccine for another devastating parasitic infection known as schistosomiasis [38], and of course our COVID-19 vaccine released for emergency use in India.

It is in that same spirit that I embarked on a parallel career of combating anti-science. My mindset for writing and publishing the book about my daughter was basically: If I don't write this or stand up for vaccines, then who will? After all, I am a vaccine scientist and have particular expertise on information that conclusively shows there is no vaccine–autism link. I also have firsthand knowledge about autism; I discussed with Rachel's medical geneticists the steps of whole exome sequencing to identify her autism gene. A downside to writing about Rachel was that it invited waves of attacks from the anti-vaccine movement. However, this also gave me experience and understanding about the tactics and approaches used by these groups. By circumstance, I became an anti-science expert.

On my early CNN and MSNBC appearances in 2020, I was among the first to call out the Trump West Wing for its use of disinformation and anti-science propaganda—not from brilliance, but because by default I had become a national expert in deciphering the rhetoric of anti-science, including especially the false narratives, which in some cases are spun around real facts. (I detail this below.) Beginning in 2021 and continuing into 2022, with 200,000 unvaccinated Americans needlessly losing their lives because of their COVID-19 vaccine defiance and refusal, I again felt an obligation to confront and combat this anti-science aggression.

Those of us in the biomedical community who feel it is a moral imperative to direct our scientific energies toward saving lives must recognize that battling anti-science is an essential element of this cause. For inspiration, I turn to the writings and thoughts of Jacob Bronowski (1908–74) [35]. Bronowski was a Polish-born mathematician and philosopher who taught mathematics in the United Kingdom and over time became a renowned public intellectual through his writings and later as the presenter of a BBC documentary (and companion book) known as *The Ascent of Man*. One of the more moving segments was filmed at Auschwitz as Bronowski spoke about a dark side of science that can be twisted for nefarious purposes. Previously, in his 1956 book, *Science and Human Values*, Bronowski wrote about the liberation of the concentration camps and Holocaust victims, as well as the aftermath of the 1945 atomic bombings of Hiroshima and Nagasaki [39]. He used these examples to develop an essential moral framework for scientists. Bronowski became one of the first "scientific humanists" in residence after the founding of the Salk Institute of Biological Sciences. When I met with Dr. Salk there in 1995, he explained to me that the rationale, during the organization's early years, for devoting a part of the institute to building bridges between the sciences and the humanities was the striking gap between the two groups. He and Bronowski believed that close intellectual contact between each discipline could significantly enrich the other. Over time, for the institute to focus on its continued excellence in biomedical sciences, this emphasis on the humanities became a lesser priority. Therefore, it remains urgent to build out meaningful programs in the humanities and public engagement at many of our major research institutions and universities. There are now some interesting and successful examples of such collaborations, including the Center for Humanities & History of Modern Biology at Cold Spring Harbor Laboratory, strong programs in the humanities at the Massachusetts Institute of Technology, and a 2018 report from the National Academy of Sciences, Engineering, and Medicine on integrating humanities and the arts into science, medicine, and engineering [40]. However, much more needs to be done.

Taking Sides

One of the most challenging aspects of confronting anti-science aggression is that those promoting its agenda have acquired wealth, power, and organization. The anti-vaccine/anti-science ecosystem now includes the most widely viewed nighttime cable news shows, far-right members of the US Congress and extremist groups, and a formidable array of contrarian intellectuals or pseudointellectuals. From my personal experience, I learned firsthand that these groups play hardball. Not only are they aggressive, but as I have tried to make clear, they do not feel compelled to be truthful. They sometimes seek to trigger waves of hate e-mails and attacks via social media.

Another challenge is the simple reality that anti-science very much runs along a partisan divide. The anti-vaccine and anti-science movements are fully enmeshed in extreme conservative or far-right politics. At times, this can include extremist politics, such as when the Proud Boys and other White nationalist groups participate in anti-vaccine rallies and messaging. Therefore, combating anti-science means it is often not possible to remain politically neutral. In my case, it is not so much that I care to enter into political disputes, but rather, what I desperately seek is to find ways to convince far-right groups to shun the anti-science element. Because anti-science is such a killer and destroyer of lives in America, my message is to say: This is not your fight. You are entitled to your conservative political views, even extremist views in many cases, but please distance yourself from the anti-science. Too often, however, my efforts to uncouple the anti-science from political extremism are interpreted as something other than my best efforts to save lives. Particularly if I say this on CNN or MSNBC, considered by the mainstream GOP and far-right groups to represent liberal views, my efforts to defeat anti-science are misinterpreted as political theater.

Complicating my urgent desire to address the partisan science-versus-anti-science divide is the fact that backing from professional scientific colleagues is often lacking. Too often they are silent or look the other way. Some of this is understandable. A study conducted by Arizona State University in 2022 found that the politicization of science has caused many scientists to withdraw from the public or greatly cut back their public engagement [41]. A component of this is attributable to increased harassment of scientists by those with political agendas, but I also believe it is because of our traditional beliefs that scientists should remain politically neutral. We are taught not to speak about Democrats and Republicans or liberals and conservatives, because we are supposed to rise above all of that. Similarly, many of the scientific academic and professional societies, as well as the national academies have historically taken the position that they must remain above the political fray. I understand the necessity of their insistence on neutrality, but this now means that many of our major scientific institutions remain on the sidelines when we need them more than ever to challenge anti-science aggression.

Given the intense politically partisan nature of the anti-science movement, the scientific community is perplexed about how to proceed. Standing up to anti-science does at times require engaging in an uncomfortable dialogue about Republicans and far-right extremist groups. My premise is that it is impossible to talk about it without talking about it, but I also concede that there is no clear road map for advancing a discussion without giving the appearance of playing politics. When I attempt to diffuse hostility by emphasizing I have no intention of interfering with my listeners' conservative or even far-right views—it is anti-science that must be eschewed at all costs because it threatens lives, including their own—the audience, more often than not, responds as though I am nevertheless engaging in a political discussion and taking sides. Recently, while giving a guest lecture in a class on medicine and ethics in the philosophy department at the University of Texas–Austin, the course director and my colleague, Professor Sahotra Sarkar, pointed out to me that remaining apolitical in the face of anti-science aggression is itself a choice not to act. Steering clear of the aggressor or tormentor is also undermining our profession by fostering an environment of fear among scientists. So I am still learning, mostly through trial and error, how to engage on an issue that is so starkly partisan without falling into the trap of becoming overtly political. Biomedical anti-science aggression from the far-right is still relatively new, so establishing the best strategy to engage and diffuse it while minimizing our direct entry into politics remains problematic.

Purveyors of Uncertainty

Still another uncomfortable but necessary discussion concerns the health freedom propagandists who seek to discredit mainstream research and raise doubts about the integrity of prominent US scientists. Before and during the pandemic I was fortunate enough to speak on several occasions with Dr. David Broniatowski. Among other things, David looks at computer algorithms to understand the vaccine hesitancy landscape, and as I pointed out earlier, he has helped lead efforts to provide information on Russian involvement and how the Putin propaganda machine systematically weaponizes communication about vaccines to create a wedge and divide our country.

In April 2022, I was on a vaccine hesitancy panel with Dr. Broniatowski at the World Vaccine Congress in Washington, DC, discussing tactics sometimes used by contrarian intellectuals to undermine public confidence in science and scientists. He described such individuals as "merchants of uncertainty," referring to their efforts to portray scientists such as Dr. Fauci and myself as individuals who espouse scientific dogma rather than scientific truths. His statement may also allude to the landmark 2010 book titled *Merchants of Doubt*, written by two science historians, Naomi Oreskes and Erik M. Conway, about how groups of contrarian scientists and intellectuals worked to cause public confusion or doubt regarding mainstream science for an array of topics—such as climate change, acid rain, the ozone layer, and tobacco—in support of the financial interests of major companies and stakeholders [42].

According to Dr. Broniatowski, the contrarians boldly state that genuine science is never certain or settled and that we can never really know what is true. Or to paraphrase David, in this postmodern era, "you have your narrative; I have mine." In this manner, almost anyone with scientific credentials can weave a false story about the dangers or lack of effectiveness of COVID-19 vaccines or other prevention measures. If then, as a major COVID-19 scientist in America, I attempt to portray the contrarian viewpoints as risible or ridiculous, I become little more than a purveyor of propaganda, while the true scientists or intellectuals are represented by those who are "simply asking questions," a common refrain from Tucker Carlson on his Fox News broadcast. David has also explained to me how this approach resembles a well-known propaganda technique designed to cause uncertainty and raise doubts about the veracity of either side of an issue.

In March 2022, an essay appeared on the Cato Institute website titled "Against Scientific Gatekeeping: Science Should Be a Profession, Not a Priesthood" [43], which also appeared in *Reason*, a libertarian magazine. The Cato Institute is a Washington, DC–based libertarian think tank that was founded by Charles Koch and Edward Crane in the 1970s [44]. The author of the essay, Dr. Jeffrey Singer, is a surgeon and senior fellow at the institute. The points he makes, together with some of the public responses to them, help us to understand some of the framing of the issues highlighted above. When hydroxychloroquine was shown to be ineffective against COVID-19 after multiple clinical trials [45–48], the FDA appropriately revoked its emergency use authorization. This left millions of emergency doses of hydroxychloroquine stockpiled and unused. Yet Dr. Singer points out, "There is a difference, however, between the claim that a drug has been proven not helpful and the weaker claim that it has not been proven helpful.... [M]any Americans believed that hydroxychloroquine's potential benefits outweighed its minimal risks. Exercising their right to self-medicate, some people infected by the coronavirus continued to take the drug" [43].

He then proceeds to his major point that "[s]cience should be a profession, not a priesthood," further arguing, "Because the internet has democratized science, the academy no longer has a monopoly on specialized information," while "search engines and the digitization of scientific literature have forever eroded their authority as gatekeepers of knowledge." Dr. Singer defends the contrarian intellectuals as experts who question "the orthodoxy," claiming they are unfairly dismissed as "fringe." Quoting President Dwight Eisenhower, Singer reminds us that "we should be alert to the ... danger that public policy could itself become captive of a scientific technological elite." He then attempts to characterize Dr. Edward Jenner's discovery of the smallpox vaccine and the firing of Dr. Ignaz Semmelweis from the Vienna General Hospital because he recommended hand washing between obstetrical deliveries as noble challengers of the orthodoxy of this medical priesthood.

Singer's essay represents an articulation of the major tenets of health freedom, although updated to incorporate the Internet into the public square. He asserts that biomedical science needs democratization and anyone can have opinions. Established scientists no longer deserve gatekeeper status. Therefore, the benefits of spectacular cures such as hydroxychloroquine or ivermectin and other unconventional practices must belong to the people, not the priesthood of scientists.

David Gorski, MD, PhD, FACS, is a distinguished cancer surgeon and physician-scientist at Wayne State University School of Medicine and the Barbara Ann Karmanos Cancer Institute. He is also a well-established vaccine advocate who is a close colleague of mine. In a responsive blog post, Gorski (writing under his pen name, Orac) explains that Singer's view of science as a religion or cult and scientists as priests is a "long-standing trope" used for decades by anti-vaccine or anti-science groups [49]. Gorski pokes holes in Singer's argument that the mainstream scientific community dismissed fringe theories because those who promote them are not adequately credentialed as epidemiologists or other kinds of biomedical scientists. He rightly points out that not belonging to the guild or "priesthood" had little if anything to do with being disregarded but rather that the ideas were not adequately supported by hard data compared with the more mainstream theories. Gorski also notes, "While it is true that sometimes people outside of a field can 'provide valuable perspectives that can be missed by those within it,' more commonly they make simple mistakes based on superficial and incomplete knowledge of the field that members of the field spot immediately" [49]. Along those lines, I'm reminded of the Cornell University astronomer Carl Sagan and his famous statement: "But the fact that some geniuses were laughed at does not imply that all who are laughed at are geniuses. They laughed at Columbus, they laughed at Fulton, they laughed at the Wright brothers. But they also laughed at Bozo the Clown" [50].

Dr. Gorski further points out that the findings of Jenner and others whom Singer mentioned were, in fact, accepted by the mainstream scientific community rather quickly once their results were found to be reproducible. He concludes: "The problem is physicians ... who don't see the problem in how that democratization has been weaponized by ideologues to oppose science-based policies against the pandemic. It's not just the pandemic, either. A similar dynamic is at work in climate science, evolution, 'integrative medicine,' and the war against women's reproductive health and the rights of LGBTQIA individuals, which science is misrepresented and misused to justify" [49].

Similar sentiments were expressed in an essay by Drs. Gorski and Gavin Yamey from Duke University titled "Covid-19 and the New Merchants of Doubt" [51], while Drs. Tara Smith and Steven Novella, writing about HIV/AIDS denialists—those who claim AIDS does not really exist or is not caused by a virus—also draw similar conclusions [52]. They further point out that the science denialists are not restricted to using the scientific literature for sharing their theories, and most scientists are not actively engaged on social media or in other public forums. In this way, the anti-scientists can gain the upper hand: "Every medical field has its legitimate controversies and complexities, and the process of science is often messy. Denial groups exploit the gap between public education and scientific reality" [52].

Setting Us Up to Fail

In the biomedical sciences, anti-science groups exploit to their advantage two key tactics that make it difficult for the scientific community to counter their influence. First, anti-science in America is currently spurred by a strong partisan divide, but the scientific professions remain committed to political neutrality. Next, health freedom propaganda often dismisses mainstream science as little more than science dogma perpetuated by high priests working at elite research universities or institutes. To make matters worse, the anti-science groups dominate the modern public square—the Internet and social media—knowing full well that our profession looks inward, seldom engages the public, and prefers journals and scientific conferences where we speak only to other scientists.

Therefore, success in combating anti-science aggression requires that we must at some level be prepared to do battle on multiple fronts. It means that at least some biomedical scientists must show a willingness to learn and practice science communication in the public marketplace. It also means that we must face the reality that neutrality favors the aggressor, and we must start feeling comfortable in the political realm. Finally, we cannot be intimidated by accusations that we are priests or gatekeepers. There is room for legitimate scientific disagreement and debate, but we must recognize that those who posit a false equivalency between ivermectin or hydroxychloroquine and lifesaving vaccines have merely weaponized science communication.

Going up against anti-science groups, especially those who profess allegiance to authoritarian pursuits or extremist politics, is daunting. We cannot expect all or even a majority of biomedical scientists to take on such a mission, or, to echo the words at the entrance to the magic theater in the Herman Hesse novel *Der Steppenwolf*, "Entrance Not for Everybody." However, we know that counteracting anti-science aggression is no abstract pursuit. Across America, anti-science attitudes have caused massive losses in human life during the second and third years of the pandemic. We should also accept that anti-vaccine activism and its deadly health freedom propaganda will not end with COVID-19. That is only the beginning; next will be working to discredit all vaccinations. Anti-science now extends to other key aspects of biomedicine, including those that depend on fetal tissue research, fundamental virology, gene editing, and many other scientific pursuits.

Nothing in life is to be feared, it is only to be understood. Now is the time to understand more, so that we may fear less.
—Marie Curie

Complaining about a problem without posing a solution is called whining.
—Theodore Roosevelt

8 | Southern Poverty Law Center for Scientists

More than 20 years ago, the author and public intellectual Malcolm Gladwell wrote about "tipping points" to explain how a major social trend gains critical mass before it spills across societies [1]. Although COVID-19 acted as the accelerant for widespread anti-vaccine aggression, its major elements were already in place by 2019. A health freedom campaign had first spread out of Texas in the 2010s and had taken hold even before the pandemic. It became a key part of the political platforms of first the Tea Party and then the far-right. Extremist elements within the conservative movement embraced the central tenets of health freedom propaganda, which included reframing vaccine refusal and defiance as a core American value that prioritized vaccine choice over public safety, touted unproven treatments, and pushed nutritional supplements to promote natural immunity as superior to supposedly toxic vaccine ingredients. All of these pieces drove us to a tipping point.

As the pandemic unfolded, these views became central to American political life on the right. Members of Congress from the House Freedom Caucus, as well as some red state legislatures and governors, openly questioned the safety or effectiveness of vaccines, while demonizing them as political instruments of control. Both Media Matters and the Center for Law and Health at ETH Zurich (considered Europe's equivalent of the Massachusetts Institute of Technology) documented how this new outlook was promulgated nightly on Fox News and other conservative outlets, becoming talking points for podcasters. Meanwhile, a group of contrarians from think tanks and universities gave it academic cover and legitimacy. Unfortunately, this challenge to medical science did not stop at vaccines. A new triumvirate of elected officials, conservative news outlets, and pseudointellectuals, as well as the courts, now voiced their disdain for masks and other public health measures. Eventually they claimed that COVID-19 was deliberately created in a laboratory supported by funds from the NIAID-NIH and began to vilify prominent US biomedical scientists. In 2022, the Florida commissioner of education, Richard Corcoran, began banning certain math textbooks on the grounds that they "incorporate prohibited topics or unsolicited strategies" [2]. The governor praised him. A nation built in part on the ingenuity and commitment of scientists and scientific institutions flipped on its side. Now science and scientists were objects of derision and in some circles considered enemies of the state.

Those tuning in nightly to Fox News or paying attention to the rhetoric of House Freedom Caucus Republicans refused COVID-19 vaccinations. When the hurricane known as the COVID-19 Delta variant reached the United States in 2021, it found lots of warm water in the form of millions of unvaccinated Americans who also considered masks and other prevention measures contemptible. Many were Republicans of lower educational attainment living in rural or suburban areas of Texas and other southern states, Appalachia, and regions of the Mountain West. An estimated 200,000 died unnecessarily because they believed more in their elected officials, Fox News, and the contrarians than they did the scientists. Anti-science aggression thus became a major killing force in the United States [3]. Many additional Americans who refused vaccinations became disabled from long COVID, or their children became orphans.

Not stopping at US borders, this lethal force moved into Canada and disrupted life there. The anti-science ecosystem now pervades western Europe, where it also links with extremist elements, just as it does in the United States. The Proud Boys march at anti-vaccine rallies in this country, as do extremists in Europe. These elements form a core of authoritarianism embedded in democratic republics, and its proponents openly admire autocratic leaders and governments in Hungary, Brazil, and elsewhere. Further fuelling the movement is a Russian propaganda machine deployed by the Putin government, which sees the benefits of destabilizing democracies through anti-vaccine and anti-science disinformation. Along with the elements just highlighted, anti-vaccine activists identified by the Center for Countering Digital Hate (CCDH) are generating additional—and lethal—anti-vaccine and anti-science content [4]. The propaganda from the West and from Russia now reaches low-income nations in Africa and Asia, where it interferes with COVID-19 vaccine uptake. A summary of this frightening new world order is presented in figure 8.1.

Tragically, this situation is not the beginning of the end, but more like the end of the beginning. Currently, little prevents this anti-science juggernaut from expanding. Although it reached critical mass during the pandemic, this movement is no longer only about COVID-19, and we should expect it will spill over to other areas, with a resulting drop in immunization rates for all childhood vaccines and interference with many other aspects of public health, including global efforts to combat HIV/AIDS, malaria, tuberculosis, and neglected tropical diseases. Anti-science has begun to contaminate other cutting-edge fields of biomedicine, including gene editing, bioinformatics, stem cell research, fetal medicine, systems biology, transplant biology, and modern neuroscience. This will only get worse.

Southern Poverty Law Center for Scientists

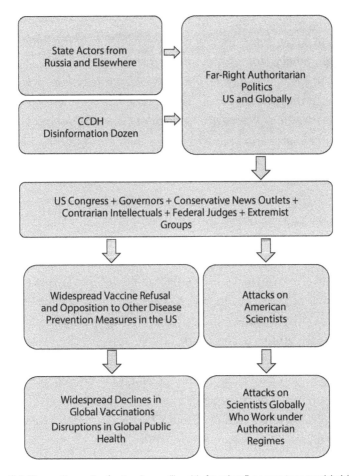

Figure 8.1. The anti-vaccine/anti-science "health freedom" ecosystem worldwide.

In the meantime, the US government response to anti-science aggression remains modest. The Department of Health and Human Services (DHHS) focuses its energies on Facebook and the social media companies, encouraging them to adjust their computer algorithms to reduce the tidal wave of disinformation. While helpful, this approach by itself does very little to stop the far-right from generating dangerous Internet content or the elected officials who campaign on their successes in attacking science and scientists. The Biden administration is concerned, but so far it has not tapped expertise outside the health sector and sought advice from cabinet departments ranging from Homeland Security to Justice and State. Similarly, the UN agencies wring their hands about the "infodemic" but do not raise this issue with authoritarian leaders in the UN Security Council or General Assembly. Halting anti-science aggression both within their borders or internationally remains a second-tier priority. Regarding GOP extremism, an umbrella under which falls this new anti-science, the Nobel laureate in economics, Paul Krugman, writes, "it cannot be appeased or compromised with. It can only be defeated" [5]. He may be correct.

We should also be concerned about additional drivers. Although most mainstream religions support vaccinations and scientific accomplishments, there are fringe elements that do not. This explains the assassinations of polio vaccinators in Afghanistan and Pakistan—both Muslim-majority nations—and the reemergence of measles among Orthodox Jewish groups in New York, New Jersey, and elsewhere in 2019. Some Evangelical Christian groups also now adopt anti-vaccine platforms [6], at times forming dangerous alliances with conservative or far-right groups to discourage vaccinations and other COVID-19 preventions. Today, Islam, Judaism, and Christianity together have more than four billion adherents. We must work with religious leaders to halt anti-science beliefs from going mainstream.

Regarding Evangelical Christianity, some have argued that anti-vaccine activism is far less tied to actual religion than it is to core values of the Middle American Radicals who feel oppressed by America's elites living in the Northeast or California [7]. Yet another term sometimes used for this disaffected group is the "New Right," whose members ("New Rightists") are not "doctrinaire libertarians" but who would rather impose their own mandates, including abortion bans, rescinding the rights of the LGBTQ community, and requiring school curricula that oppose elements of critical race theories [8, 9]. These radical groups rail against what they term "woke lunacy" [7]. Perhaps in their view, vaccines and vaccinations somehow became part of the canon of wokeness, which is a reason I believe some public figures seek to discredit vaccines. In this case, we need to work harder to remind Middle America or what some call the "flyover nation" that vaccines and other products derived from cutting-edge science are not *woke*, but lifesaving interventions. We must help everyone understand that the United States is a nation built on science and technology, and that preserving its greatness depends on embracing scientific principles as a fundamental American value.

Still another concern is the fact that large numbers of healthcare professionals shunned COVID-19 vaccinations during the pandemic, including many from conservative or religious groups, according to Tim Callaghan and his colleagues [10]. Anti-vaccine activism among the health professions represents another ominous trend. Then, in 2020, Renée DiResta from the Stanford Internet Observatory raised yet another concern [11]. Artificial intelligence (AI) now generates disinformation at an accelerated pace. Her warning, published in the *Atlantic* with the title "The Supply of Disinformation Will Soon Be Infinite," suggests that anti-science aggression may soon expand its firepower capabilities. A key point to remember is that anti-science is no longer a theoretical construct. It already kills many people and causes permanent injury in countless others.

A Larger War on Science

Up until the mid-twentieth century, there was no obvious anti-science contingent in the Republican Party. At its start, the first Republican president, Abraham Lincoln, launched the National Academy of Sciences. The Eisenhower administration created the National Aeronautics and Space Administration. However, in his 2005 book, *The Republican War on Science* [12], the journalist Chris Mooney benchmarks the rise of anti-intellectualism and distrust of academics to the failed presidential bid of Senator Barry Goldwater in 1964. Such attitudes even prompted the formation of a group known as Scientists and Engineers for Johnson-Humphrey, who attacked Goldwater for his reckless language concerning the use of nuclear weapons. Later, President Richard Nixon dissolved the office of the White House science advisor (subsequently restored under Presidents Gerald Ford and Jimmy Carter) because he considered the scientific community generally opposed to his efforts to promote antiballistic missiles or a supersonic transport high-speed passenger jet. Similarly, President Ronald Reagan was slow to accept HIV/AIDS as a public health crisis or promote the use of condoms or other AIDS prevention measures, even to the point of forbidding his surgeon general, Dr. C. Everett Koop, from speaking out on these issues. Koop in his memoir stated, "I have never understood why these peculiar restraints were placed on me" [13]. The Reagan White House also promoted the teaching of creationism in schools and banned federal support for fetal tissue transplantation research.

Some of these activities continued under President George H. W. Bush, although the senior Bush also elevated the importance of the president's science advisor when he designated Yale University physics professor D. Allan Bromley as assistant to the president. However, henceforward the GOP's support for many scientific activities waned even further. Mooney points out that Rep. Newt Gingrich (R-Georgia), in his role as Speaker of the House during the 1990s, began cultivating contrarian scientists to have them testify to Congress. This ushered in a period in which science was politicized to support the business community or other national interests. The George W. Bush White House promoted efforts to counter scientific consensus views in a diverse range of areas, including climate change, environmental protection, stem cell research, and alternative medicine practices. It is notable that in the almost 20 years between 2001 and 2019 (well before COVID-19), red states and counties had a consistently higher mortality rate from a variety of conditions, including heart disease, cancer, cerebrovascular illness, and chronic lung disease [14]. Undoubtedly the higher rates of poverty and comorbidities contributed to this situation, but we must also consider the role played by a lower public acceptance of scientific medicine in these areas.

In fairness, the same George W. Bush administration also championed many global health initiatives, including the President's Emergency Plan for AIDS Relief and the President's Malaria Initiative. I personally worked with members of the Bush White House and conservative members of Congress, including then Sen. Sam Brownback (R-Kansas), to support mass treatments for neglected tropical diseases [15]. Now more than one billion people annually receive treatments for these conditions. Even then, in the early years of the 2000s, it was a straightforward proposition to work with both sides of the political aisle to promote global health legislation. I remember when Sen. Brownback invited me to speak to his prayer breakfast group, and afterward, I would go across the way to speak with liberal Democrats about the same issue. This was not even considered unusual. It was understood and accepted that bipartisanship was essential to accomplish important tasks. Politics is now more polarized than ever before in modern times. Today, I could not imagine transitioning from conservative to liberal members of Congress with the same ease and fluidity.

Climategate and Coronagate

In the years following the Nixon, Reagan, and two Bush administrations, right-wing anti-science tendencies expanded and became more ferocious. Some conservatives eventually embraced attacks on individual scientists. This was first apparent in climate science. My colleague Dr. Michael E. Mann, a distinguished climate scientist and geophysicist at the University of Pennsylvania, endured public attacks in what became known as "climategate." In 2009, climate-change denialists hacked the e-mails of Dr. Mann and his colleagues, including Professor Phil Jones at East Anglia University, claiming they had found evidence that these and other scientists manipulated their data [16]. In addition to unjust allegations that the climate scientists had conspired to make up their findings regarding global warming, they received threats that resemble the ones I now receive regarding vaccines. For example, an e-mail to one of the East Anglia University faculty read: "Just a quick note to encourage you to shoot yourself in the head." Also: "Don't waste any more time. Do it today. It is truly the greatest contribution to mankind that you will ever make" [17]. The 2008 Republican vice-presidential nominee and former Alaska governor, Sarah Palin, penned an opinion piece in the *Washington Post* that opened with the following statements: "With the publication of damaging e-mails from a climate research center in Britain, the radical environmental movement appears to face a tipping point. The revelation of appalling actions by so-called climate change experts allows the American public to finally understand the concerns so many of us have articulated on this issue" [18]. Ultimately, Mann and the other climate scientists involved were totally cleared of any wrongdoing, but not before they had to endure painful federal scientific misconduct investigations [19]. These activities ushered in a new era, in which not only was the science under attack by conservative groups or the far-right, but also the scientists themselves. When he was asked by *Science* magazine about the attacks on biomedical scientists during the COVID-19 pandemic, Mann responded, "We're here; we feel your pain" [20].

The Trump White House continued its attacks on climate science and scientists, pulling out of the Paris climate accords and recruiting into government scientists who held contrarian or fringe views on the role of human activity in global warming [21]. I have previously highlighted the failings of the Trump administration in launching a COVID-19 response in the United States during the early months of the pandemic [22], and these were summarized in 2022 by the House Select Committee on the Coronavirus Crisis [23].

A similar level of aggression and ferocity was directed at virologists conducting basic research on coronaviruses. Coronaviruses that resemble the SARS-2 have been recovered from bats and other mammals across Southeast Asia [24, 25], and a strong body of evidence supports the emergence of COVID-19 among humans in central China following transmission from exotic animals (such as raccoon dogs and pangolins) through contact in the wet markets [26, 27]. Such evidence did not halt accusations by members of Congress against several prominent coronavirus researchers, including Drs. Peter Daszak at the EcoHealth Alliance and Ralph Baric at the University of North Carolina. Their allegations assert either that the virus was engineered or created artificially in a laboratory in the United States and later sent to China, or that American scientists supported dangerous research conducted at the Wuhan Institute of Virology, located in the same city where COVID-19 may have emerged in wet markets [28]. Others claim that the SARS 2 coronavirus somehow leaked from the Wuhan Institute. Because of my own experiences, I began reaching out to Peter Daszak by phone to offer my support or just to serve as a helpful friend and listener. Peter's situation was reported by the prominent science journalist and writer Jon Cohen in his article for *Science* titled "Prophet in Purgatory":

> Daszak's emails, tweets, letters, journal articles, and media interviews have been scrutinized; he has received blistering criticism in Congress, on social media, and in major news outlets; he has been accused of conflicts of interest, a lack of transparency, being a China apologist, and conducting reckless experiments. He has received death threats, including a letter holding white powder resembling anthrax, and journalists have staked out his home to shoot photos and videos. Two high-profile commissions to study the pandemic's origin have collapsed in part because he was a member. [28]

Indeed, during our calls, Peter confirmed his distress and expressed concerns for the safety of his family. What also came through in our conversations was his righteous indignation. He became a scientist to help humankind, only to be vilified as an enemy of the state. For me as well, this aspect of the situation is especially demoralizing. We became scientists to help the nation and the world; as I have explained to Peter on several occasions, we are the true patriots, not the phony ones who attack us.

Foremost among the attacks against Dr. Daszak are claims that he received funds from NIAID-NIH to perform gain of function (GOF) research. For RNA viruses (like coronaviruses), GOF research involves the insertion of specific pieces of RNA into the virus genome to increase either the transmissibility or virulence of the virus, meaning greater severity of illness. In the case of coronaviruses, this includes insertion of small sequences of RNA that encode amino acids that render the surface spike protein of the virus more susceptible to a host cellular furin proteolytic enzyme. Such furin cleavage sites help to make SARS-2 more infectious or transmissible than its SARS-1 predecessor that emerged in 2002 [29]. However, as I have pointed out repeatedly in my interviews and in efforts to defend Peter Daszak, the simple truth is that furin cleavage sites are found in most subfamilies of coronaviruses [30, 31]. Even distinguished scientists who initially saw furin cleavage sites as a possible signature of human manipulation, subsequently backed off their claims [32]. Ultimately, several prominent biomedical scientists—including the virologists Robert Garry (Tulane University), Angela Rasmussen (University of Saskatchewan), Michael Worobey (University of Arizona), and the infectious disease and global health physician Dr. Gerald Keusch (Boston University)—wrote well-supported papers to debunk either the GOF claims or theories that somehow the SARS-2 coronavirus escaped from a laboratory [27, 33–36].

Despite the overwhelming evidence pointing to the natural origins of COVID-19, the finger-pointing related to GOF (or, alternatively, theories postulating a subsequent lab leak) continues against Daszak. In April 2022, two prominent Republican House members, Reps. Steve Scalise (R-Louisiana) and James Comer (R-Kentucky), asked the DHHS secretary to investigate Daszak and the EcoHealth Alliance for unethical conduct and to bar both him and the organization from receiving future federal grants. Such actions would likely end Peter's career as a scientist and effectively terminate the EcoHealth Alliance.

They also represent a backdoor approach by the far-right to discredit NIAID-NIH and Dr. Anthony Fauci, whose institute supported the research of both Drs. Daszak and Baric. Fox News anchors and multiple members of Congress, including Rep. Comer, Rep. Matt Gaetz (R-Florida), and Sen. Rand Paul (R-Kentucky), have made outrageous and false claims, or they have insinuated that Dr. Fauci was responsible for creating the COVID-19 virus or starting the epidemic [37]. In 2022, prior to the November midterm elections, Sen. Paul made public statements indicating he would investigate Dr. Fauci if the Republicans ever gain a majority in the Senate [38]. A follow-up 35-page report by the Republican Senate staff ignored most of the actual science and previous committee investigations from the World Health Organization and US intelligence agencies, choosing instead to tout lab leak and other conspiracies [39]. The hashtag #FauciLiedPeopleDied occasionally trends on social media.

As Dr. David Gorski points out, it is not surprising that contrarian intellectuals or far-right think tanks piled on to these accusations [40], but then, in an unusual twist, several unrelated activist groups joined in. For example, organizations opposed to genetically modified food or agricultural products, which in some cases may also have ties to anti-vaccine groups, have called for bans against any GOF research or have demanded investigations into alleged NIH-supported lab leaks [41, 42], as have also some animal rights activists [43]. Some of these same groups even tried to draw me into GOF or lab leak accusations by falsely asserting that our coronavirus vaccine development efforts somehow supported GOF-related work.

The attacks and threats against the community of COVID-19 scientists won't abate anytime soon and will likely continue well after the pandemic ends. An ominous blog called *What Covid Crimes Will Vic-*

tims Not Forgive? and published on the libertarian Brownstone Institute's website warns:

> The Covid response can and undoubtedly will be portrayed in future years as the product of criminal negligence.... Such a thing can get ugly. Once a population is truly convinced they have been betrayed by an elite that has both money and status (read: things to lose), all gloves are off. We are then in similar historical circumstances as those in which Germany found itself in the 1920s, where a belief spread in the idea that Germany had lost the Great War due to betrayal by socialists and Jews.... Just how powerful this story will turn out to be is difficult to predict, but what we can predict is who can be counted upon to champion it most vociferously: the businesspeople who irrecoverably lost their positions due to the Covid lockdowns and other restrictions, the young and single who for similar reasons lost the best years of their lives, and those who believe the vaccines did them and their children permanent damage. That alliance—forged in the fires of lasting hurt to human well-being—could produce a formidable adversary against the culpable Covid elites. [44]

Seeking Help

As I began reaching out to Michael Mann, Peter Daszak, and others under threat from anti-science groups, it became clear that certain themes were common to our journeys [45]. We recognize how anti-science aggression is deepening and widening. It was also becoming more menacing, as evidenced by the tone and details of the e-mail and social media threats—and in some cases, even physical stalkings. Another element was that the threats included specific efforts to end our scientific careers through outreach to our university administrators or fellow colleagues, as well as the agencies that fund our work. Even when the assertions have no basis in reality, the simple fact of knowing that certain people or groups seek to discredit your professional activities is unnerving. This is especially true for scientists, because our entire identity is often linked to our discoveries.

An especially chilling realization was that the attacks against us were a part of something much bigger. As I explained earlier, attacking science itself rarely suffices in a rising authoritarian regime, whose leaders soon find it necessary to go after individual scientists. We had become enemies of the state. Most of the major scientific and professional societies offer little solace because of their commitment to political neutrality. This dark period of anti-science makes clear that these organizations must undergo reform. However, it is unlikely that this would be a quick process. In the meantime, the scientific community currently works under the threat of powerful and authoritarian segments of a society that dominates many portions of the cable news outlets, the Internet, and the halls of Congress. These realities underlie my plea to fellow scientists to expand our level of public engagement and science communication. This includes communicating in new and different ways, such as appealing to emotion and core human values, but also not being afraid to take a political stand. Political neutrality, while desirable, may not always be possible when anti-science aggression displays such a strong partisan divide, and so many Americans lose their lives in our red states. It is essential to remember that scientists have not politicized vaccines and COVID-19 prevention measures. Instead, this was done by elected officials, conservative news outlets, the courts, and contrarians. Our job is to say: This makes no sense, and it is destroying lives. Our obligation is to uncouple the anti-science movement from political extremism on the right.

To enlist the help of colleagues and to seek their advice, I reached out to two informal groups. One included regular e-mails or Zoom chats with Peter Daszak and two esteemed colleagues, Drs. Gerald Keusch and Rich Roberts. Dr. Keusch was former director of the NIH

Fogarty International Center, and Rich Roberts is a Nobel laureate for his work in eukaryotic RNA expression who helped to create New England BioLabs. While our group was originally organized to help advise Peter Daszak in managing the unfair and unjust assaults on his character and organization, in time, as I became a target for very different reasons—my staunch support of vaccines—I found the calls personally useful for comparing notes and even risk-management strategies for potential physical confrontations. I also joined a second informal group formed by the renowned University of Minnesota epidemiologist Dr. Mike Osterholm. Mike convened a distinguished group of physicians and scientists that included two former FDA commissioners—Drs. Margaret Hamburg and Stephen Hahn—as well as Drs. Eric Topol (Scripps Research Translational Institute), Bruce Gellin (Rockefeller Foundation), Penny Heaton (Gates Foundation and later Johnson & Johnson), and Ruth Berkelman (CDC). In time, all these individuals became friends who helped to guide me through anti-science aggression during the time of the pandemic. Finally, because of the anti-Semitic leanings of many of the anti-science attacks, B'nai B'rith International and the Anti-Defamation League have offered me assistance and valuable advice.

These experiences have led me to think about less-fortunate colleagues who do not have access to prominent scientists for advice and help when targeted by anti-science groups. Some university or academic medical center offices of communications, or their general counsels, employ trained professionals to help students, postdoctoral fellows, resident physicians, and faculty navigate the complexities countering anti-science aggression. However, the level of support varies a lot by institution. Some discourage their young scientists from engaging the public because it potentially opens the institution up to public criticism. All too often, scientists receive the message (either subliminally or overtly) that they speak out at their own risk.

The reality is that the needs of scientists under threat from anti-science forces are pervasive. Among the many kinds of help we require in these circumstances is counsel on managing threats delivered via e-mail or social media, including coordinated bullying campaigns, as well as attacks by the conservative media. It is not unusual for even senior scientists to face reputational loss, or potential loss of employment, as a result of such threats. Still another type of intimidation is known as "legal thuggery," in which anti-science groups or individuals use letters written by attorneys to threaten defamation or libel suits. Such letters seek to frighten scientists into silence. Unfortunately, this technique is sometimes quite effective. As another example of legal thuggery, Peter Daszak tells me that both he and the EcoHealth Alliance have faced lawsuits alleging that they ignited the COVID-19 pandemic. Even though these charges have no basis, such lawsuits require hiring law firms, paying legal fees, and dealing with the emotional stress that accompanies these activities. Other prominent US biomedical scientists have also needed legal aid in response to anti-science aggression. Given that anti-science aggression may accelerate through the use of artificial intelligence and other forces, it is especially important that the scientific community prepare and respond.

A New Type of Support

In an ideal setting, one of the current scientific or professional societies or the national academies would step up to create organized and comprehensive assistance to scientists in distress. Some pieces are already in place. For instance, the Committee of Concerned Scientists is an important international organization that tracks human rights abuses against scientists in countries such as China, Turkey, and Iran [46], but it was not necessarily created for the complex attacks now occurring from the authoritarian right in the United States.

In response to the attacks on Michael Mann and other climate scientists, a Climate Science Legal Defense Fund was created as a 501(c)3 non-profit to help with legal costs or defense against lawsuits [47]. Today, it supports scientists who are under threat or silenced and also works to strengthen legal protections for scientists while promoting the integrity of climate science. It offers free legal aid, educates scientists about their rights and responsibilities, shares strategies with attorneys, and publicizes attacks on climate science and scientists. Therefore, one idea is to establish a similar defense fund for biomedical or COVID-19 scientists. On a 2022 Zoom call with an individual from the Climate Science Legal Defense Fund, together with representatives from the Union of Concerned Scientists, I discussed this possibility. The Union of Concerned Scientists is a nonprofit advocacy group founded in 1969 by the students and faculty of the Massachusetts Institute of Technology for reshaping government science policies to address environmental and nonmilitary applications [48]. This organization first reached out to me following my *Boston Globe* opinion piece on biomedical scientists under threat [49].

Among the ideas I floated on the call, and one I also wrote about for the Federation of American Societies for Experimental Biology, the largest society of experimental biologists in America, was to create an entity that resembled the Southern Poverty Law Center [50], but for biomedical scientists [51]. The Southern Poverty Law Center was created in 1971 to combat racism and promote civil rights through a legal framework. The organization is also a clearinghouse for collecting information about hate groups, working with law enforcement to protect people of color and other victims of discrimination. A good strategy might be to establish a similar organization, this one committed to the protections of biomedical scientists and willing to champion our rights, especially when we are under attack by extremist elements. Such an organization should be one that can respond quickly to the needs of US scientists across the country, with deep familiarity with the various types of threats launched by far-right and other anti-science groups. The type of risk-management help should range from legal advice to managing online threats and even assistance with law enforcement. Another opportunity might be to expand the COVID-19 Hate Crimes Act, adopted in 2021 to protect Asian-Americans against political violence [52], to protect American scientists as well. Given the increase of anti-science aggression in the United States and its incorporation into mainstream conservative politics, such remedies could address the urgent needs of the mainstream scientific community.

Response from the US Government

The fact that the DHHS and US surgeon general have responded at all and that they now work with the major social media platforms is a positive development and one that should continue to be encouraged. However, these actions do not address those generating the content from the far-right, the role of the disinformation dozen in monetizing the Internet, or the Russian government's weaponized health communication. Given the 20 years of relative neglect by the US government in tackling anti-science aggression, I believe we must realize that this issue goes way beyond the health sector. We need input from other branches of the federal government such as the Departments of Homeland Security, Commerce, Justice—and even State, given the Russian involvement. We must seek ways to demonetize the use of the Internet by the disinformation dozen or halt the anti-science aggression emanating from Fox News and elected officials, but in ways that do not violate the Bill of Rights or the US Constitution. Although the health sector may not know what can and should be done to address anti-science aggression, there are those who do and who could come to the table with experiences that taught them how to combat global terrorism, cyberattacks, and nuclear proliferation. We must learn from them. Along those lines, the White House should consider establishing an interagency task force to examine such possibilities and to make recommendations for action to slow the progression of anti-science. And given the globalization of the anti-science ecosystem, especially in Canada and Europe, we can also seek help and input from international organizations such as the North Atlantic Treaty Organization or the United Nations. Until

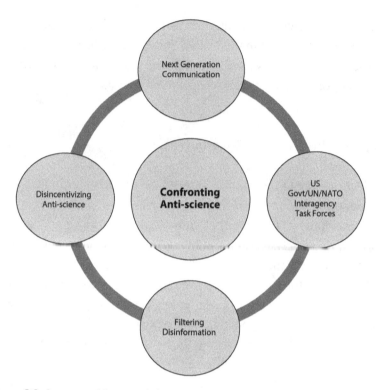

Figure 8.2. A proposed framework for combating anti-science aggression.

now, such agencies have been employed to combat more conventional and globalizing threats. Anti-science aggression now warrants this level of engagement and a counterresponse (fig. 8.2).

As someone who personally endures attacks from conservative news outlets and US elected officials, I speak from experience when I say that it is increasingly apparent that we need a clearinghouse and organization to specifically support or defend scientists in America.

We are a nation built on our great research universities and institutes, so we therefore must recognize the contributions of scientists to the history of the United States. The only way America can remain a role model for the rest of the world is to solidly back our scientists and keep them free from the threats and pressures of political extremist groups. Otherwise, we risk losing our best minds to other endeavors. It is essential that young people growing up in the United States recognize the scientific profession as important, and one worth defending. In turn, those who come to America for scientific training must see it as a nation that honors its scientists and is perceived as an international role model. In parallel, we must engage the White House and international organizations such as NATO or the UN to lead a response that no longer relies exclusively on the health sector. We need the advice of experts who understand big-picture global threats such as terrorism, cyberattacks, or nuclear proliferation. Finally, we need a new framework for science communication—one that is unafraid of political engagement or confrontation and is willing to shepherd and train the next generation of biomedical scientists.

Biomedical science in America is under threat, and this has deadly consequences. Anti-science aggression is a leading killer of unvaccinated adults who have been rendered unnecessarily vulnerable to COVID-19. Because of anti-science and its expanded organization, funding, and political power, responding to the next pandemic will be even more challenging. However, even today, many biomedical scientists live in an environment of fear or uncertainty just by pursuing their daily activities. The COVID-19 pandemic is only the beginning. Anti-science has become a new normal that threatens American democratic principles and our way of life. The sad reality is that it is now apparent that outside groups are not coming to our rescue and that the scientific community must help itself—especially the biomedical scientists currently in jeopardy from far-right, authoritarian, and extremist groups. A timely, strong, and robust response from the scientific community is essential. The future of scientific inquiry in the United States and globally depends on how we regroup and act.

Literature Cited

1 | An Army of Patriots Turns against the Scientists

1. Pinna M, Picard L, Goessmann C. Cable news and COVID-19 vaccine uptake. Sci Rep 12(1) (October 7, 2022):16804, doi: 10.1038/s41598-022-20350-0.
2. Hotez PJ. Will anti-vaccine activism in the USA reverse global goals? Nat Rev Immunol 22 (2022): 525–26, doi: 10.1038/s41577-022-00770-9.
3. Lee SK, Sun J, Jang S, Connelly S. Misinformation of COVID-19 vaccines and vaccine hesitancy. Sci Rep 12(1) (August 11, 2022):13681, doi: 10.1038/s41598-022-17430-6.
4. Hotez PJ. How the anti-vaxxers are winning. New York Times, February 8, 2017, https://www.nytimes.com/2017/02/08/opinion/how-the-anti-vaxxers-are-winning.html.
5. Hotez PJ. Vaccines Did Not Cause Rachel's Autism: My Journey as a Vaccine Scientist, Pediatrician, and Autism Dad. Baltimore, MD: Johns Hopkins University Press, 2018.
6. Wakefield AJ, Murch SH, Anthony A, Linnell J, Casson DM, Malik M, Berelowitz M, Dhillon AP, Thomson MA, Harvey P, Valentine A, Davies SE,

Literature Cited

Walker-Smith JA. Ileal-lymphoid-nodular hyperplasia, non-specific colitis, and pervasive developmental disorder in children. Lancet 351(9103) (February 28, 1998): 637–41, doi: 10.1016/s0140-6736(97)11096-0. Retraction in: Lancet 375(9713) (February 6, 2010): 445. Erratum in: Lancet 363(9411) (March 6, 2004): 750.

7. Olive JK, Hotez PJ, Damania A, Nolan MS. The state of the antivaccine movement in the United States: A focused examination of nonmedical exemptions in states and counties. PLOS Med 15(6) (2018): e1002578, https://doi.org/10.1371/journal.pmed.1002578.
8. Hotez PJ. The antiscience movement is escalating, going global and killing thousands. Sci Am, March 29, 2021, https://www.scientificamerican.com/article/the-antiscience-movement-is-escalating-going-global-and-killing-thousands.
9. Hotez PJ, Nuzhath T, Colwell B. Combating vaccine hesitancy and other 21st century social determinants in the global fight against measles. Curr Opin Virol 41 (April 2020): 1–7, doi: 10.1016/j.coviro.2020.01.001.
10. Tanne JH. First polio case in decades reported in the Americas. BMJ, July 25, 2022, doi: 10.1136/bmj.o1864.
11. Center for Countering Digital Hate. The disinformation dozen: Why platforms must act on twelve leading online anti-vaxxers, March 2021, https://www.counterhate.com/disinformationdozen.
12. Ahmed I. Dismantling the anti-vaxx industry. Nat Med 27(3) (March 2021): 366, doi: 10.1038/s41591-021-01260-6.
13. Associated Press. "This is a disinformation industry": Meet the media startups making big money on vaccine conspiracies. Fortune, May 14, 2021, https://fortune.com/2021/05/14/disinformation-media-vaccine-covid19.
14. Cockerell I. Anti-vaxxers make up to $1.1 billion for social media companies. Coda, June 2, 2021, https://www.codastory.com/waronscience/social-media-profit-pandemic-antivax.
15. Essley Whyte L. Spreading vaccine fears: And cashing in. Center for Public Integrity, June 8, 2021, https://publicintegrity.org/health/coronavirus-and-inequality/spreading-fears-cashing-in-anti-vaccine.
16. Milmo D. Anti-vaxxers making "at least $2.5m" a year from publishing on Substack. Guardian, January 27, 2022, https://www.theguardian.com/technology/2022/jan/27/anti-vaxxers-making-at-least-25m-a-year-from-publishing-on-substack.

17. Hotez PJ. The medical biochemistry of poverty and neglect. Mol Med 20 Suppl 1 (December 16, 2014): S31–36, doi: 10.2119/molmed.2014.00169.
18. Hotez PJ, Bottazzi ME. A COVID vaccine for all. Sci Am, December 30, 2021, https://www.scientificamerican.com/article/a-covid-vaccine-for-all.
19. Texas Children's Hospital. Texas-developed patent-free COVID-19 vaccine technology receives emergency use authorization in Indonesia. Press release, https://www.texaschildrens.org/about-us/news/releases/texas-developed-patent-free-covid-19-vaccine-technology-receives-emergency-use-authorization.
20. Ida Nurcahyani, RS. Bio Farma's IndoVac vaccine obtains BPJPH's halal certificate. Antara News, October 6, 2022, https://en.antaranews.com/news/253461/bio-farmas-indovac-vaccine-obtains-bpjphs-halal-certificate.
21. Rosenfeld JA, Xiao R, Bekheirnia MR, Kanani F, Parker MJ, Koenig MK, van Haeringen A, Ruivenkamp C, Rosmaninho-Salgado J, Almeida PM, Sá J, Pinto Basto J, Palen E, Oetjens KF, Burrage LC, Xia F, Liu P, Eng CM; Undiagnosed Diseases Network, Yang Y, Posey JE, Lee BH. Heterozygous variants in SPTBN1 cause intellectual disability and autism. Am J Med Genet A 185(7) (July 2021): 2037–45, doi: 10.1002/ajmg.a.62201.
22. Hotez PJ, Molyneux DH, Fenwick A, Kumaresan J, Sachs SE, Sachs JD, Savioli L. Control of neglected tropical diseases. N Engl J Med 357(10) (September 6, 2007):1018–27, doi: 10.1056/NEJMra064142.
23. Thielking M, Branswell H. CDC expects "community spread" of coronavirus, as top official warns disruptions could be "severe." Stat News, February 25, 2020, https://www.statnews.com/2020/02/25/cdc-expects-community-spread-of-coronavirus-as-top-official-warns-disruptions-could-be-severe.
24. Hotez PJ. Texas and its measles epidemics. PLOS Med 13(10) (October 25, 2016): e1002153, https://doi.org/10.1371/journal.pmed.1002153.
25. Bissada M. Anti-vaccine mandate protest: RFK Jr., Proud Boys, and holocaust imagery. Forbes, January 23, 2022, https://www.forbes.com/sites/masonbissada/2022/01/23/anti-vaccine-mandate-protest-rfk-jr-proud-boys-and-holocaust-imagery/?sh=5137bdc67942.
26. O'Grady C. In the line of fire. Science 375(6587) (March 24, 2022): 1338–43, https://www.science.org/content/article/overwhelmed-hate-covid-19-scientists-face-avalanche-abuse-survey-shows.
27. Nogrady B. Scientists under attack. Nature 598 (October 14, 2021): 250–52.

Literature Cited

28. Mooney C. The Republican War on Science. New York: Basic Books, 2006.
29. Leonhardt D. Red Covid (Morning Newsletter). New York Times, September 27 (updated October 1), 2021, https://www.nytimes.com/2021/09/27/briefing/covid-red-states-vaccinations.html.
30. Leonhardt D. U.S. Covid deaths get even redder (Morning Newsletter). New York Times, November 8 (updated November 24), 2021, https://www.nytimes.com/2021/11/08/briefing/covid-death-toll-red-america.html.
31. Leonhardt D. Red Covid, an update. New York Times, February 18, 2022, https://www.nytimes.com/2022/02/18/briefing/red-covid-partisan-deaths-vaccines.html.
32. Gaba's data and Hamel's comment are cited in: Wood D. Pro-Trump counties now have far higher COVID death rates: Misinformation is to blame. National Public Radio, December 5, 2021, https://www.npr.org/sections/health-shots/2021/12/05/1059828993/data-vaccine-misinformation-trump-counties-covid-death-rate.
33. Gaba C. Challenge accepted: The elephant in the room, now age-adjusted. ACASignups.net, February 2, 2022, https://acasignups.net/22/02/03/challenge-accepted-elephant-room-now-age-adjusted.
34. Gaba C. Weekly update: County-level #COVID19 vaccination levels by 2020 partisan lean. ACASignups.net, February 14, 2022, https://acasignups.net/22/02/14/weekly-update-county-level-covid19-vaccination-levels-2020-partisan-lean.
35. Gaba C, Stokes A. "Non-COVID" excess death rates ran 21x higher in reddest counties than bluest in 2021. ACA Signups, May 5, 2022, https://acasignups.net/22/05/09/exclusive-non-covid-excess-death-rates-ran-21x-higher-reddest-counties-bluest-2021.
36. Chase W, Owens C. The pandemic has been deadlier in red states. Axios, March 25, 2022, https://www.axios.com/coronavirus-pandemics-politics-masks-vaccines-deaths-2644f22b-fa77-4e34-a6fe-8b492172dff2.html.
37. Bump P. How much of right-wing opposition to vaccination was Fox News's fault? Washington Post, October 10, 2022, https://www.washingtonpost.com/politics/2022/10/10/vaccines-coronavirus-fox-news.
38. Hotez P. COVID vaccines: Time to confront anti-vax aggression. Nature 592(7856) (April 2021): 661, doi: 10.1038/d41586-021-01084-x.
39. World Health Organization. COVID-19 pandemic leads to major backsliding on childhood vaccinations, new WHO, UNICEF data shows. Press release, July 15, 2021, https://www.who.int/news/item/15-07-2021

-covid-19-pandemic-leads-to-major-backsliding-on-childhood-vaccinations-new-who-unicef-data-shows.

40. Shet A, Carr K, Danovaro-Holliday MC, Sodha SV, Prosperi C, Wunderlich J, Wonodi C, Reynolds HW, Mirza I, Gacic-Dobo M, O'Brien KL, Lindstrand A. Impact of the SARS-CoV-2 pandemic on routine immunisation services: Evidence of disruption and recovery from 170 countries and territories. Lancet Glob Health 10(2) (February 2022): e186–e194, doi: 10.1016/S2214-109X(21)00512-X.

41. Patel Murthy B, Zell E, Kirtland K, Jones-Jack N, Harris L, Sprague C, Schultz J, Le Q, Bramer CA, Kuramoto S, Cheng I, Woinarowicz M, Robison S, McHugh A, Schauer S, Gibbs-Scharf L. Impact of the COVID-19 pandemic on administration of selected routine childhood and adolescent vaccinations—10 U.S. jurisdictions, March–September 2020. MMWR Morb Mortal Wkly Rep 70(23) (June 11, 2021): 840–45, doi: 10.15585/mmwr.mm7023a2.

42. St. George D. Student vaccinations slowed during Covid: Can schools catch them up? Washington Post, August 26, 2022, https://www.washingtonpost.com/education/2022/08/26/student-vaccination-measles-polio-school.

43. Orac. Antivaxxers rejoice over the spillover of distrust of COVID-19 vaccines to all vaccines. Respectful Insolence, August 22, 2022, https://www.respectfulinsolence.com/2022/08/22/antivaxxers-rejoice-over-the-spillover-of-distrust-of-covid-19-vaccines-to-all-vaccines/.

44. Wilkinson E. Is anti-vaccine sentiment affecting routine childhood immunisations? BMJ 376 (February 10, 2022): o360, doi: 10.1136/bmj.o360.

45. Butler K. Anti-vaxxers have a new obsession: Keeping your kids from seeing their doctors. Mother Jones, February 25, 2022, https://www.motherjones.com/politics/2022/02/anti-vaxxers-have-a-new-obsession-keeping-your-kids-from-seeing-their-doctors.

46. Global Burden of Disease Study 2019, Institute for Health Metrics and Evaluation, https://ghdx.healthdata.org/gbd-results-tool.

47. Hotez PJ. Preventing the Next Pandemic: Vaccine Diplomacy in a Time of Anti-science. Baltimore, MD: Johns Hopkins University Press, 2021.

48. Uwishema O, Elebesunu EE, Bouaddi O, Kapoor A, Akhtar S, Effiong FB, Chaudhary A, Onyeaka H. Poliomyelitis amidst the COVID-19 pandemic in Africa: Efforts, challenges and recommendations. Clin Epidemiol Glob Health 16 (July–Aug 2022): 101073, doi: 10.1016/j.cegh.2022.101073.

49. Anthes E. Polio may have been spreading in New York since April. New York Times, August 16, 2022, https://www.nytimes.com/2022/08/16/health/polio-new-york.html.
50. Link-Gelles R, Lutterloh E, Ruppert PS, Backenson PB, St. George K, Rosenberg ES, Anderson BJ, Fuschino M, Popowich M, Punjabi C, Souto M, McKay K, Rulli S, Insaf T, Hill D, Kumar J, Gelman I, Jorba J, Ng TFF, Gerloff N, Masters NB, Lopez A, Dooling K, Stokley S, Kidd S, Oberste MS, Routh J; 2022 U.S. Poliovirus Response Team. Public health response to a case of paralytic poliomyelitis in an unvaccinated person and detection of poliovirus in wastewater—New York, June–August 2022. Am J Transplant. 22(10) (October 2022): 2470–74, doi: 10.1111/ajt.16677.
51. Ryerson AB, Lang D, Alazawi MA, Neyra M, Hill DT, St George K, Fuschino M, Lutterloh E, Backenson B, Rulli S, Ruppert PS, Lawler J, McGraw N, Knecht A, Gelman I, Zucker JR, Omoregie E, Kidd S, Sugerman DE, Jorba J, Gerloff N, Ng TFF, Lopez A, Masters NB, Leung J, Burns CC, Routh J, Bialek SR, Oberste MS, Rosenberg ES; 2022 U.S. Poliovirus Response Team. Wastewater testing and detection of poliovirus type 2 genetically linked to virus isolated from a paralytic polio case—New York, March 9–October 11, 2022. MMWR Morb Mortal Wkly Rep. 71(44) (November 4, 2022): 1418–24, doi: 10.15585/mmwr.mm7144e2.

2 | Health Freedom Propaganda in America

1. Mitropoulos A, Hartung K, Said S. Even on their death beds, some COVID-19 patients in Idaho still reject vaccination. ABC News, September 11, 2021, https://abcnews.go.com/Health/death-beds-covid-19-patients-idaho-reject-vaccination/story?id=79941785.
2. Farhi P. Four conservative radio talk-show hosts bashed coronavirus vaccines: Then they got sick. Washington Post, September 1, 2021, https://www.washingtonpost.com/lifestyle/media/conservative-talk-radio-covid-deaths/2021/08/31/a912a89c-0a66-11ec-aea1-42a8138f132a_story.html.
3. Stunson M. Christian radio host who asked if vaccine is form of government control dies of COVID. Charlotte Observer, August 18, 2021, https://www.charlotteobserver.com/news/coronavirus/article253569969.html.
4. Ecarma C. Anti-vax radio hosts keep dying from COVID. Vanity Fair, September 3, 2021, https://www.vanityfair.com/news/2021/09/anti-vax-radio-hosts-dying-covid.

5. Vorel J. At least 7 anti-vaccine, anti-mask conservative activists have died of COVID-19 in recent weeks. Paste magazine, September 16, 2021, https://www.pastemagazine.com/politics/coronavirus/covid-anti-vaccine-anti-mask-radio-hosts-dead-tucker-carlson-fox-news/.
6. Petrizzo Z, Venarchik A. Anti-vax radio host who got COVID at QAnon-friendly conference dies. Daily Beast, January 6, 2022, https://www.thedailybeast.com/covid-infected-radio-host-douglas-kuzma-dies-after-qanon-friendly-conference-with-baseless-anthrax-rumors.
7. Sommer W. QAnon star who said only "idiots" get vax dies of COVID. Daily Beast, January 7, 2022, https://www.thedailybeast.com/qanon-star-cirsten-weldon-who-said-only-idiots-get-vaccinated-dies-of-covid.
8. Treisman R. COVID was again the leading cause of death among U.S. law enforcement in 2021. National Public Radio, January 12, 2022, https://www.npr.org/2022/01/12/1072411820/law-enforcement-deaths-2021-covid.
9. https://twitter.com/jasonrantz/status/1450086763569958916, accessed May 12, 2022.
10. Connor T, Olding R, Bolies C. Anti-vax cop who told guv to "kiss my ass" dies after battle with COVID. Daily Beast, January 29, 2022, https://www.thedailybeast.com/robert-lamay-anti-vax-washington-state-trooper-who-railed-at-jay-inslee-dies-after-covid-battle.
11. Fenton T. Fox News under fire for failing to report on trooper's death from Covid after they aired his anti-vax stance. Yahoo News, February 1, 2022, https://www.yahoo.com/video/fox-news-under-fire-failing-141559550.html.
12. Health Freedom for All Act, H.R. 5471, 117th Congress (September 30, 2021), https://www.congress.gov/bill/117th-congress/house-bill/5471/text.
13. White House. Briefing room fact sheet: Biden administration announces details of two major vaccination policies, November 4, 2021, https://www.whitehouse.gov/briefing-room/statements-releases/2021/11/04/fact-sheet-biden-administration-announces-details-of-two-major-vaccination-policies.
14. Hals T. Republican governors lead attack on Biden vaccine mandate. Reuters, November 5, 2021, https://www.reuters.com/world/us/republican-governors-lead-attack-biden-vaccine-mandate-2021-11-05.
15. Mettler K, Johnson L, Moyer JW, Contrera J, Davies E, Silverman E, Hermann P, Jamison P. Anti-vaccine activists march in D.C.—a city that man-

dates coronavirus vaccination—to protest mandates. Washington Post, January 23, 2022, https://www.washingtonpost.com/dc-md-va/2022/01/23/dc-anti-vaccine-rally-mandates-protest.
16. Devine C, Griffin D. Leaders of the anti-vaccine movement used "stop the steal" crusade to advance their own conspiracy theories. CNN, February 5, 2021, https://www.cnn.com/2021/02/04/politics/anti-vaxxers-stop-the-steal-invs/index.html.
17. Wire SD. Beverly Hills anti-vaccine doctor pleads guilty in Jan. 6 Capitol riot case. Los Angeles Times, March 3, 2022, https://www.latimes.com/politics/story/2022-03-03/beverly-hills-anti-vax-doctor-pleads-guilty-in-jan-6-capitol-riot-case.
18. Orso, RJ. Feds charge an N.J. anti-vax activist and a correctional officer in Capitol riot. Philadelphia Inquirer, January 22, 2021, https://www.inquirer.com/news/new-jersey/stephanie-hazelton-capitol-insurrection-medford-new-jersey-rioters-20210122.html.
19. Ford M. In a 6–3 ruling, the Supreme Court upholds the Covid pandemic. New Republic, January 13, 2022, https://newrepublic.com/article/165018/covid-vaccine-mandate-struck-down.
20. Hotez PJ. America's deadly flirtation with antiscience and the medical freedom movement. J Clin Invest 131(7) (April 1, 2021): e149072, doi: 10.1172/JCI149072.
21. Grossman LA. The origins of American health libertarianism. Yale J Health Policy Law Ethics 13(1) (2013): 76–134.
22. Hotez PJ. Loss of laboratory instruction in American medical schools: Erosion of Flexner's view of "scientific medical education." Am J Med Sci 325(1) (January 2003): 10–14, doi: 10.1097/00000441-00301000-00003.
23. Flexner A. Medical Education in the United States and Canada. Classics of Medicine Library, 1990.
24. Wolfe RM, Sharp LK. Anti-vaccinationists past and present. BMJ 325(7361) (2002): 430–32.
25. National Health Federation. Missions and values, https://thenhf.com/about-nhf/our-mission-values.
26. Lyons RD. Rightists are linked to laetrile's lobby. New York Times, July 5, 1977, https://www.nytimes.com/1977/07/05/archives/rightists-are-linked-to-laetriles-lobby-but-backers-of-purported.html.
27. Health Freedom Protection Act, H.R. 4282, 109th Congress (2005), https://www.congress.gov/bill/109th-congress/house-bill/4282.

28. Satija N, Sun L. A major funder of the anti-vaccine movement has made millions selling natural health products. Washington Post, December 20, 2019, https://www.washingtonpost.com/investigations/2019/10/15/fdc01078-c29c-11e9-b5e4-54aa56d5b7ce_story.html.
29. Wakefield AJ, Murch SH, Anthony A, Linnell J, Casson DM, Malik M, Berelowitz M, Dhillon AP, Thomson MA, Harvey P, Valentine A, Davies SE, Walker-Smith JA. Ileal-lymphoid-nodular hyperplasia, non-specific colitis, and pervasive developmental disorder in children. Lancet 351(9103) (February 28, 1998): 637–41, doi: 10.1016/s0140-6736(97)11096-0. Retraction in: Lancet 375(9713) (February 6, 2010): 445. Erratum in: Lancet 363(9411) (March 6, 2004): 750.
30. Hotez PJ. Vaccines Did Not Cause Rachel's Autism: My Journey as a Vaccine Scientist, Pediatrician, and Autism Dad. Baltimore, MD: Johns Hopkins University Press, 2021.
31. Mohanty S, et al. Experiences with medical exemptions after a change in vaccine exemption policy in California. Pediatrics 142(5) (2018): e20181051.
32. Smith TC. Vaccine rejection and hesitancy: A review and call to action. Open Forum Infect Dis 4(3) (2017): ofx146.
33. Lakshmanan R, Sabo J. Lessons from the front line: Advocating for vaccines policies at the Texas capitol during turbulent times. J Applied Res Child 10(2) (2019): article 6.
34. Texans for Vaccine Choice. Protecting your right to choose, https://texansforvaccinechoice.com, accessed November 18, 2022.
35. Sun LH. Trump energizes the antivaccine movement in Texas. Washington Post, February 20, 2017, https://www.washingtonpost.com/national/health-science/trump-energizes-the-anti-vaccine-movement-in-texas/2017/02/20/795bd3ae-ef08-11e6-b4ff-ac2cf509efe5_story.html.
36. Wheat A. Tim Dunn is against the nanny state—except when it might help his oil business. Texas Monthly, April 23, 2020, https://www.texasmonthly.com/news-politics/tim-dunn-against-nanny-state-except-help-oil-business.
37. Novack S. How anti-vaxxers are injecting themselves into the Texas Republican primaries. Texas Observer, February 28, 2018, https://www.texasobserver.org/anti-vaxxers-injecting-texas-republican-primaries.
38. Hotez PJ. Texas and its measles epidemics. PLOS Med 13(10) (2016): e1002153, https://doi.org/10.1371/journal.pmed.1002153.

Literature Cited

39. Hotez PJ. How the anti-vaxxers are winning. New York Times, February 8, 2017, https://www.nytimes.com/2017/02/08/opinion/how-the-anti-vaxxers-are-winning.html.
40. Texas Department of State Health Services. Conscientious exemptions data–vaccine coverage levels, https://www.dshs.texas.gov/immunize/coverage/Conscientious-Exemptions-Data.shtm, accessed January 30, 2022.
41. Olive JK, Hotez PJ, Damania A, Nolan MS. The state of the antivaccine movement in the United States: A focused examination of nonmedical exemptions in states and counties. PLOS Med 15(6) (2018): e1002578, doi: 10.1371/journal.pmed.1002578.
42. Sun L. U.S. measles cases surge to second-highest level in nearly two decades. Washington Post, April 1, 2019, https://www.washingtonpost.com/health/2019/04/01/us-measles-cases-surge-second-highest-level-nearly-two-decades.
43. Bailey JM. A California legislator ended his efforts to pass a vaccine mandate for students. New York Times, April 15, 2022, https://www.nytimes.com/2022/04/15/us/a-california-legislator-ended-his-efforts-to-pass-a-vaccine-mandate-for-students.html.
44. Brumfiel G. Inside the growing alliance between anti-vaccine activists and pro-Trump Republicans. National Public Radio, December 6, 2021, https://www.npr.org/2021/12/06/1057344561/anti-vaccine-activists-political-conference-trump-republicans.
45. Hotez P. The U.S. anti-vax movement is contaminating Canada (commentary). Global News, September 14, 2021, https://globalnews.ca/news/8187962/the-u-s-anti-vax-movement-is-contaminating-canada.
46. Lau Y. Canada's "Freedom Convoy" has shut down Ottawa and is blocking $500 million daily in cross-border trade: Here's where things stand. Fortune, February 16, 2022, https://fortune.com/2022/02/16/canada-freedom-convoy-protests-whats-happening.
47. Liles J. Swastikas and Confederate flags seen at Canada's "freedom convoy." Snopes, February 17, 2022, https://www.snopes.com/news/2022/02/17/swastikas-canada-freedom-convoy.
48. Lardner R, Smith MR, Swenson A. How American right-wing funding for Canadian trucker protests could sway U.S. politics. PBS News Hour and Associated Press, February 17, 2022, https://www.pbs.org/newshour/world/how-american-right-wing-funding-for-canadian-trucker-protests-could-sway-u-s-politics.

49. https://twitter.com/laurenboebert/status/1491488032914087938, accessed May 12, 2022.
50. Ngo M, Bednar A, Ray E. Led by truckers, hundreds of vehicles protesting Covid mandates encircle Washington. New York Times, March 6, 2022, https://www.nytimes.com/2022/03/06/us/trucker-convoy-dc-beltway.html.
51. Hotez P. The antiscience movement is escalating, going global and killing thousands. Sci Am, March 29, 2021, https://www.scientificamerican.com/article/the-antiscience-movement-is-escalating-going-global-and-killing-thousands.
52. Picheta R. Europe's loud, rule-breaking unvaccinated minority are falling out of society. CNN, January 16, 2022, https://www.cnn.com/2022/01/16/europe/europe-covid-unvaccinated-society-cmd-intl/index.html.
53. Khazan O. What's really behind global vaccine hesitancy. Atlantic, December 6, 2021, https://www.theatlantic.com/politics/archive/2021/12/which-countries-have-most-anti-vaxxers/620901.
54. Amiel S. Who are France's anti-vaccine rule protesters and what do they want? EuroNews, June 8, 2021, https://www.euronews.com/2021/07/26/who-are-france-s-anti-vaccine-rule-protesters-and-what-do-they-want.
55. Protesters hit French streets to fight new vaccine pass. AFP and Arab News, updated January 16, 2022, https://www.arabnews.com/node/2005241/world.
56. Previdelli A. EXPLAINED: Who are MFG—Austria's vaccine-sceptic party? Local At, February 28, 2022, local.at/20220228/explained-who-are-mfg-austrias-vaccine-sceptic-party.
57. Shields M, Chopra T. Austria mourns suicide of doctor targeted by anti-COVID vaccine campaigners. Reuters, July 30, 2022, https://www.reuters.com/world/europe/austria-mourns-suicide-doctor-targetted-by-anti-vaccine-campaigners-2022-07-30.
58. Verweij H, Deutsch A, Potter M, Smith A. Dutch police disperse anti-lockdown protesters in Amsterdam. Reuters, January 2, 2022, https://www.reuters.com/business/media-telecom/dutch-police-disperse-thousands-protesting-against-lockdown-measures-2022-01-02.
59. Marone F. Far right extremism and anti-vaccine conspiracy: A case from Italy. Italian Institute for International Political Studies, October 21, 2021, https://www.ispionline.it/en/pubblicazione/far-right-extremism-and-anti-vaccine-conspiracy-case-italy-32078.

Literature Cited

60. Broderick R. Italy's anti-vaccination movement is militant and dangerous. Foreign Policy, November 13, 2021, https://foreignpolicy.com/2021/11/13/italy-anti-vaccination-movement-militant-dangerous.
61. Exit Staff. Albanian childhood disease vaccination rates plummet. Exit News, August 3, 2022, https://exit.al/en/albanian-childhood-disease-vaccination-rates-plummet.
62. Brezar A. How a fake jabs probe highlights Greece's deep vaccine skepticism. Euro News my.europe, November 20, 2021, https://www.euronews.com/my-europe/2021/11/20/how-a-fake-jabs-probe-highlights-greece-s-deep-vaccine-scepticism.
63. Dotto C, Cubbon S. Disinformation exports: How foreign anti-vaccine narratives reached West African communities online. First Draft, June 23, 2021, https://firstdraftnews.org/long-form-article/foreign-anti-vaccine-disinformation-reaches-west-africa.
64. Davis TP Jr, Yimam AK, Kalam MA, Tolossa AD, Kanwagi R, Bauler S, Kulathungam L, Larson H. Behavioural determinants of COVID-19-vaccine acceptance in rural areas of six lower- and middle-income countries. Vaccines (Basel) 10(2) (January 29, 2022): 214, doi: 10.3390/vaccines10020214.
65. Hotez PJ. Will anti-vaccine activism in the USA reverse global goals? Nat Rev Immunol (August 1, 2022): 1–2, doi: 10.1038/s41577-022-00770-9.
66. Jarry J. Infertility: A diabolical agenda is anti-vaxx sleight-of-hand propaganda. McGill University Office of Science and Society, July 12, 2022, https://www.mcgill.ca/oss/article/covid-19-critical-thinking-pseudoscience/infertility-diabolical-agenda-anti-vaxx-sleight-hand-propaganda.
67. Onyango E. Stephen Karanja: Kenyan anti-vaccine doctor dies from Covid-19. BBC News, April 30, 2021, https://www.bbc.com/news/world-africa-56922517.
68. Butler K, Wadekar N. How American influencers build a world wide web of vaccine disinformation. Mother Jones, June 2, 2022, https://www.motherjones.com/politics/2022/06/how-american-influencers-built-a-world-wide-web-of-vaccine-disinformation.
69. Vaccination Demand Observatory, https://dashboard.thevdo.org.
70. United Nations. UN condemns brutal killing of eight polio workers in Afghanistan. February 24, 2022, https://news.un.org/en/story/2022/02/1112612.

3 | Red COVID

1. Mueller B, Lutz E. U.S. has far higher Covid death rate than other wealthy countries. New York Times, February 1, 2022, https://www.nytimes.com/interactive/2022/02/01/science/covid-deaths-united-states.html.
2. COVID-19 Excess Mortality Collaborators. Estimating excess mortality due to the COVID-19 pandemic: A systematic analysis of COVID-19-related mortality, 2020–21. Lancet 399(10334) (April 16, 2022): 1513–36, doi: 10.1016/S0140-6736(21)02796-3.
3. Institute of Health Metrics and Evaluation, University of Washington. COVID-19 Projections, https://covid19.healthdata.org/global, accessed March 2022. The data can be accessed by moving the cursor to the specific dates.
4. Center for Systems Science and Engineering, Johns Hopkins University COVID-19 dashboard, https://gisanddata.maps.arcgis.com/apps/dashboards/bda7594740fd40299423467b48e9ecf6.
5. White House. Briefing room fact sheet: President Biden to announce all Americans to be eligible for vaccinations by May 1, puts the nation on a path to get closer to normal by July 4th, March 11, 2022, https://www.whitehouse.gov/briefing-room/statements-releases/2021/03/11/fact-sheet-president-biden-to-announce-all-americans-to-be-eligible-for-vaccinations-by-may-1-puts-the-nation-on-a-path-to-get-closer-to-normal-by-july-4th.
6. Institute for Government. Coronavirus vaccine rollout, https://www.instituteforgovernment.org.uk/explainers/coronavirus-vaccine-rollout.
7. Hotez PJ. The great Texas COVID tragedy. PLOS Glob Public Health 2(10) (October 2022): e0001173, https://doi.org/10.1371/journal.pgph.0001173.
8. Brooks Harper K. Unvaccinated Texans make up vast majority of COVID-19 cases and deaths this year, new state data shows. Texas Tribune, November 8, 2021, https://www.texastribune.org/2021/11/08/texas-coronavirus-deaths-vaccinated.
9. Johnson AG, Amin AB, Ali AR, et al. COVID-19 incidence and death rates among unvaccinated and fully vaccinated adults with and without booster doses during periods of Delta and Omicron variant emergence—25 U.S. jurisdictions, April 4–December 25, 2021. MMWR Morb Mortal Wkly Rep 71 (2022): 132–38, https://www.cdc.gov/mmwr/volumes/71/wr/mm7104e2.htm.

Literature Cited

10. Lucar J, Wingler MJB, Cretella DA, Ward LM, Sims Gomillia CE, Chamberlain N, Shimose LA, Brock JB, Harvey J, Wilhelm A, Majors LT, Jeter JB, Bueno MX, Albrecht S, Navalkele B, Mena LA, Parham J. Epidemiology, clinical features, and outcomes of hospitalized adults with COVID-19: Early experience from an academic medical center in Mississippi. South Med J 114(3) (March 2021): 144–49, doi: 10.14423/SMJ.0000000000001222.
11. Amin K, Ortaliza J, Cox C, Michaud J, Kates J. COVID-19 mortality preventable by vaccines. Peterson-KFF Health System Tracker, April 21, 2022, https://www.healthsystemtracker.org/brief/covid19-and-other-leading-causes-of-death-in-the-us.
12. Gaba C. My own crude estimate: Vaccine refusal has likely killed 180K–235K Americans to date. ACASignups.net, March 10, 2022, https://acasignups.net/22/03/10/my-own-crude-estimate-vaccine-refusal-has-likely-killed-180k-235k-americans-date.
13. Jia KM, Hanage WP, Lipsitch M, Swerdlow D. Excess COVID-19 associated deaths among the unvaccinated population ≥18 years old in the US, May 30–December 4, 2021. MedR$_x$iv, doi: https://doi.org/10.1101/2022.02.10.22270823.
14. Ndugga N, Hill L, Artiga S, Haldar. Latest data on COVID-19 vaccinations by race/ethnicity. Kaiser Family Foundation, February 2, 2022, https://www.kff.org/coronavirus-covid-19/issue-brief/latest-data-on-covid-19-vaccinations-by-race-ethnicity.
15. Rockefeller Foundation. Creating vaccine equity, https://www.rockefellerfoundation.org/covid-19-response/creating-vaccine-equity.
16. Hood Medicine. HoodMed chats, https://www.hoodmedicine.org/hood-chats.
17. Callaghan T, Moghtaderi A, Lueck JA, Hotez P, Strych U, Dor A, Fowler EF, Motta M. Correlates and disparities of intention to vaccinate against COVID-19. Soc Sci Med 272 (March 2021): 113638, doi: 10.1016/j.socscimed.2020.113638.
18. Kirzinger A, Kearney A, Hamel L, Brodie M. KFF COVID-19 vaccine monitor: The increasing importance of partisanship in predicting COVID-19 vaccination status. KFF, November 16, 2021, https://www.kff.org/coronavirus-covid-19/poll-finding/importance-of-partisanship-predicting-vaccination-status.
19. Todd C, Murray M, Kamisar B. NBC News poll shows demographic break-

down of the vaccinated in the U.S. NBC News, August 24, 2021, https://www.nbcnews.com/politics/meet-the-press/nbc-news-poll-shows-demographic-breakdown-vaccinated-u-s-n1277514.
20. Tyson A, Funk C. Increasing public criticism, confusion over COVID-19 response in US. Pew Research Center, February 9, 2022, https://www.pewresearch.org/science/2022/02/09/increasing-public-criticism-confusion-over-covid-19-response-in-u-s.
21. Scherer LD, McPhetres J, Pennycook G, Kempe A, Allen LA, Knoepke CE, Tate CE, Matlock DD. Who is susceptible to online health misinformation? A test of four psychosocial hypotheses. Health Psychol 40(4) (April 2021): 274–84, doi: 10.1037/hea0000978.
22. Hotez PJ, Jackson Lee S. US Gulf Coast states: The rise of neglected tropical diseases in "flyover nation." PLOS Negl Trop Dis 11(11) (2017): e0005744, https://doi.org/10.1371/journal.pntd.0005744.
23. Leonhardt D. Red Covid (Morning Newsletter). New York Times, September 27 (updated October 1), 2021, https://www.nytimes.com/2021/09/27/briefing/covid-red-states-vaccinations.html.
24. Leonhardt D. U.S. Covid deaths get even redder (Morning Newsletter). New York Times, November 8 (updated November 24), 2021, https://www.nytimes.com/2021/11/08/briefing/covid-death-toll-red-america.html.
25. Gaba's data and Hamel's comment are cited in: Wood D. Pro-Trump counties now have far higher COVID death rates: Misinformation is to blame. National Public Radio, December 5, 2021, https://www.npr.org/sections/health-shots/2021/12/05/1059828993/data-vaccine-misinformation-trump-counties-covid-death-rate.
26. Gaba C. Challenge accepted: The elephant in the room, now age-adjusted. ACASignups.net, February 2, 2022, https://acasignups.net/22/02/03/challenge-accepted-elephant-room-now-age-adjusted.
27. Gaba C. Weekly update: County-level #COVID19 vaccination levels by 2020 partisan lean. ACASignups.net, February 14, 2022, https://acasignups.net/22/02/14/weekly-update-county-level-covid19-vaccination-levels-2020-partisan-lean.
28. Gaba C, Stokes A. "Non-COVID" excess death rates ran 21x higher in reddest counties than bluest in 2021. ACA Signups, May 5, 2022, https://acasignups.net/22/05/09/exclusive-non-covid-excess-death-rates-ran-21x-higher-reddest-counties-bluest-2021.

Literature Cited

29. VanLaeys M. COVID is killing Republican base. Madison.com, January 15, 2022, https://madison.com/opinion/letters/covid-is-killing-republican-base----mark-vanlaeys/article_ab3870f9-a3f0-575c-bdc4-ba69d035cf84.html.
30. Douthat R. How Republicans failed the unvaccinated. New York Times, April 6, 2022, https://www.nytimes.com/2022/04/06/opinion/covid-vaccine-republicans.html.
31. Chase W, Owens C. The pandemic has been deadlier in red states. Axios, March 25, 2022, https://www.axios.com/coronavirus-pandemics-politics-masks-vaccines-deaths-2644f22b-fa77-4e34-a6fe-8b492172dff2.html.
32. Leonhardt D. Red Covid, an update. New York Times, February 18, 2022, https://www.nytimes.com/2022/02/18/briefing/red-covid-partisan-deaths-vaccines.html.
33. Nirappil F, Keathing D. Covid deaths no longer overwhelmingly among the unvaccinated as toll on elderly grows. Washington Post, April 29, 2022, https://www.washingtonpost.com/health/2022/04/29/covid-deaths-unvaccinated-boosters.
34. Link-Gelles R, Levy ME, Gaglani M, et al. Effectiveness of 2, 3, and 4 COVID-19 mRNA vaccine doses among immunocompetent adults during periods when SARS-CoV-2 Omicron BA.1 and BA.2/BA.2.12.1 sublineages predominated—VISION Network, 10 states, December 2021–June 2022. MMWR Morb Mortal Wkly Rep 71 (2022): 931–39, doi: http://dx.doi.org/10.15585/mmwr.mm7129e1.
35. Sexton A. New Hampshire House approves bill to block federal vaccine mandates for county nursing homes, state hospital. WMUR 9, February 17, 2022, https://www.wmur.com/article/new-hampshire-house-bill-federal-vaccine-mandate-nursing-homes-state-hospital-21722/39125883.
36. Gaba C. New NBER study confirms pretty much everything I (& others) have been saying for the past year & a half. ACA Signups, October 3, 2022, https://acasignups.net/22/10/03/new-nber-study-confirms-pretty-much-everything-i-others-have-been-saying-past-year-half.
37. Wallace J, Goldsmith-Pinkham P, Schwartz JL. Excess death rates for Republicans and Democrats during the COVID-19 pandemic. National Bureau of Economic Research Working Paper 30512, September 2022, doi: 10.3386/w30512.
38. Tracking coronavirus in Texas: Latest map and cases count. New York Times, updated daily, https://www.nytimes.com/interactive/2021/us/texas-covid-cases.html.

39. Hammons C. The Great Depression and WWII: 1930–1945. Texas Our Texas, https://texasourtexas.texaspbs.org/the-eras-of-texas/great-depression-ww2.
40. Burnett J. The tempest at Galveston: "We knew there was a storm coming, but we had no idea." National Public Radio, November 30, 2017, https://www.npr.org/2017/11/30/566950355/the-tempest-at-galveston-we-knew-there-was-a-storm-coming-but-we-had-no-idea.
41. Levin J, Bradshaw M. Determinants of COVID-19 skepticism and SARS-CoV-2 vaccine hesitancy: Findings from a national population survey of U.S. adults. BMC Public Health 22(1) (May 25, 2022): 1047, doi: 10.1186/s12889-022-13477-2.
42. Judi J. The return of the Middle American Radical. National Journal and Carnegie Endowment for International Peace, October 2, 2015, https://carnegieendowment.org/2015/10/02/return-of-middle-american-radical-pub-61534.
43. Commissioners of the Lancet Commission on Vaccine Refusal, Acceptance, and Demand in the USA. Announcing the Lancet Commission on Vaccine Refusal, Acceptance, and Demand in the USA. Lancet 397(10280) (March 27, 2021): 1165–67, doi: 10.1016/S0140-6736(21)00372-X.
44. Sharfstein JM, Callaghan T, Carpiano RM, Sgaier SK, Brewer NT, Galvani AP, Lakshmanan R, McFadden SM, Reiss DR, Salmon DA, Hotez PJ; Lancet Commission on Vaccine Refusal, Acceptance, and Demand in the USA. Uncoupling vaccination from politics: A call to action. Lancet 398(10307) (October 2, 2021): 1211–12, doi: 10.1016/S0140-6736(21)02099-7.
45. CDC National Center for Health Statistics. Leading causes of death, https://www.cdc.gov/nchs/fastats/leading-causes-of-death.htm.

4 | An Anti-science Political Ecosystem

1. Kelly M. Surgeon general calls out platforms over COVID-19 misinformation. Verge, July 15, 2021, https://www.theverge.com/2021/7/15/22578431/white-house-surgeon-general-facebook-youtube-twitter-covid-19-misinformation-vaccines.
2. Hotez PJ. How the anti-vaxxers are winning. New York Times, February 8, 2017, https://www.nytimes.com/2017/02/08/opinion/how-the-anti-vaxxers-are-winning.html.
3. Office of the Surgeon General. US Department of Health and Human Services. Confronting health misinformation: The surgeon general's advisory

Literature Cited

on building a healthy information environment, July 15, 2021, https://www.hhs.gov/surgeongeneral/reports-and-publications/health-misinformation/index.html.

4. Office of the Surgeon General Press Office. US Department of Health and Human Services. U.S. surgeon general issues advisory during COVID-19 vaccination push warning American public about threat of health misinformation, July 15, 2021, https://www.hhs.gov/about/news/2021/07/15/us-surgeon-general-issues-advisory-during-covid-19-vaccination-push-warning-american.html.
5. Hotez P, Batista C, Ergonul O, Figueroa JP, Gilbert S, Gursel M, Hassanain M, Kang G, Kim JH, Lall B, Larson H, Naniche D, Sheahan T, Shoham S, Wilder-Smith A, Strub-Wourgaft N, Yadav P, Bottazzi ME. Correcting COVID-19 vaccine misinformation: Lancet Commission on COVID-19 Vaccines and Therapeutics Task Force members. EClinicalMedicine 33 (March 2021): 100780, doi: 10.1016/j.eclinm.2021.100780.
6. Hotez P, Bottazzi ME. Five myths about coronavirus vaccines: No, the mRNA vaccines don't change your DNA. Washington Post, March 19, 2021, https://www.washingtonpost.com/outlook/five-myths/five-myths-about-coronavirus-vaccines/2021/03/19/0f186f8e-881f-11eb-82bc-e58213caa38e_story.html.
7. Center for Countering Digital Hate. The disinformation dozen: Why platforms must act on twelve leading online anti-vaxxers, March 21, 2021, https://www.counterhate.com/disinformationdozen.
8. DiResta R. How Amazon's algorithms curated a dystopian bookstore. Wired, March 5, 2019, https://www.wired.com/story/amazon-and-the-spread-of-health-misinformation.
9. Gray L. Dr. Peter Hotez's battle against the "anti-science confederacy" is a lifetime in the making. Houston Chronicle, March 8, 2021, https://www.houstonchronicle.com/life/article/Peter-Hotez-fight-for-science-COVID-houston-texas-15982489.php.
10. Shadow State. Robert F. Kennedy Jr. warns about Fauci, Bill Gates and coronavirus vaccines. Frank Report, May 9, 2020, https://frankreport.com/2020/05/09/robert-f-kennedy-jr-warns-about-fauci-bill-gates-and-coronavirus-vaccines.
11. Facebook, https://www.facebook.com/resultswithmary/videos/10163288539190142, accessed May 1, 2022.
12. Guccione Jr B. The outsider. Spin, January 17, 2022, https://www.spin.com/2022/01/robert-f-kennedy-jr-interview-2022.

13. The Virality Project, Cryst E, DiResta R, Meyersohn L. Memes, magnets and microchips: Narrative dynamics around COVID-19 vaccines. Stanford Digital Repository, https://purl.stanford.edu/mx395xj8490.
14. Reyna VF. A scientific theory of gist communication and misinformation resistance, with implications for health, education, and policy. Proc Natl Acad Sci 118(15) (April 15, 2021): e1912441117, doi: 10.1073/pnas.1912441117.
15. Shoaib A. White nationalist Nick Fuentes banned from annual gatherings of top US conservatives but other far-right extremists were welcomed. Business Insider, July 11, 2021, https://www.businessinsider.com/dallas-nick-fuentes-banned-far-right-extremists-roam-cpac-2021-7.
16. Holmes D. They clapped for death at CPAC. Esquire. July 12, 2021, https://www.esquire.com/news-politics/a37001629/cpac-vaccination-goal-biden-miss-clap.
17. Castronuovo C. Cawthorn: Biden door-to-door vaccine strategy could be used to "take" guns, Bibles. The Hill, July 9, 2021, https://thehill.com/homenews/house/562372-cawthorn-biden-door-to-door-vaccine-strategy-could-be-used-to-take-guns-bibles.
18. Smith A. Conservative hostility to Biden vaccine push surges with Covid cases on the rise. NBC News, July 19, 2021, https://www.nbcnews.com/politics/politics-news/conservative-hostility-biden-vaccine-push-surges-covid-cases-rise-n1273692.
19. LeBlanc P. Marjorie Taylor Greene compares Biden vaccine push to Nazi-era "brown shirts" weeks after apologizing for Holocaust comments. CNN, July 7, 2021, https://www.cnn.com/2021/07/07/politics/marjorie-taylor-greene-brown-shirts-vaccine/index.html.
20. Nguyen T. "Idiots," "anarchists," and "assholes": John Boehner unloads on Republicans in post-retirement interview. Vanity Fair, October 30, 2017, https://www.vanityfair.com/news/2017/10/john-boehner-on-republican-party.
21. Altschuler GC. The clear and present danger of Jim Jordan & Co. The Hill, December 12, 2021, https://thehill.com/opinion/campaign/585413-the-clear-and-present-danger-of-jim-jordan-co.
22. Henderson A. "A race to the bottom": House GOP slammed for "really disgraceful" anti-vaccine tweet. Salon, January 1, 2022, https://www.salon.com/2022/01/01/a-race-to-the-bottom-slammed-for-really-disgraceful-anti-vaccine-tweet.
23. Solender A. GOP Rep. Brooks pushes anti-vaccine talking points in letter

Literature Cited

to Biden. Forbes, July 19, 2021, https://www.forbes.com/sites/andrew solender/2021/07/19/gop-rep-brooks-pushes-anti-vaccine-talking-points-in-letter-to-biden/?sh=7e171e2354be.

24. Grayer A, Fox L, Fortinsky S. Not all Republicans are embracing McConnell's vaccine push: Read what some had to say when asked this week. CNN, July 22, 2021, https://www.cnn.com/2021/07/22/politics/house-republicans-vaccination-rates/index.html.
25. Paul Gosar, Twitter, September 28, 2022, https://twitter.com/repgosar/status/1575160963321319424.
26. Lybrand H, Subramani T. Fact-checking Sen. Ron Johnson's anti-vaccine misinformation. CNN, May 10, 2021, https://www.cnn.com/2021/05/07/politics/ron-johnson-vaccine-misinformation-fact-check/index.html.
27. Redman H. In interview, Sen. Johnson says it "may be true" that COVID vaccines cause AIDS. Wisconsin Examiner, May 3, 2022, https://wisconsinexaminer.com/brief/in-interview-sen-johnson-says-it-may-be-true-that-covid-vaccines-cause-aids.
28. Dickeson K. Sen. Ron Johnson hosts COVID-19 panel focusing on vaccine skepticism: Local doctors respond. NBC 26 Green Bay (Wisconsin), January 24, 2022, https://www.nbc26.com/news/local-news/sen-ron-johnson-hosts-covid-19-panel-focusing-on-vaccine-skepticism-local-doctors-respond.
29. Stolberg SG, Anti-vaccine doctor has been invited to testify before Senate committee. New York Times, December 6, 2020, updated April 26, 2021, https://www.nytimes.com/2020/12/06/us/politics/anti-vax-scientist-senate-hearing.html.
30. Vakil C. Woman fails to prove the COVID-19 vaccine made her magnetic during Ohio House hearing. The Hill, June 12, 2021, https://thehill.com/policy/healthcare/558144-woman-fails-to-prove-the-covid-19-vaccine-made-her-magnetic-during-ohio.
31. Karlin S. Anti-vaccine debate blows up in House committee as Louisiana struggles with COVID immunizations. Advocate (Baton Rouge), May 17, 2021, https://www.theadvocate.com/baton_rouge/news/politics/legislature/article_10ce704a-b744-11eb-b1be-1790b534d2a9.html.
32. Cillizza C. What Rand Paul gets wrong on vaccines. CNN Politics, May 24, 2021, https://www.cnn.com/2021/05/24/politics/rand-paul-vaccines-covid-19/index.html.
33. McCaul M, et al. McCaul urges President Biden to partner with Texas

Children's Hospital and Baylor College of Medicine on vaccine diplomacy, February 1, 2022, https://mccaul.house.gov/media-center/press-releases/mccaul-urges-president-biden-to-partner-with-texas-childrens-hospital.
34. CBS News. Florida will be "the first state to officially recommend against the COVID-19 vaccines for healthy children," state surgeon general declares, March 8, 2022, https://www.cbsnews.com/news/florida-recommend-no-covid-19-vaccines-healthy-children-surgeon-general-joseph-ladapo.
35. Florida Health Press Release. State Surgeon General Dr. Joseph A. Ladapo issues new mRNA COVID-19 Vaccine Guidance, October 7, 2022, https://www.floridahealth.gov/newsroom/2022/10/20220512-guidance-mrna-covid19-vaccine.pr.html.
36. Dyer O. Covid 19: Florida surgeon general says state will be first not to recommend vaccination for children. BMJ 376 (March 9, 2022): o622, doi: 10.1136/bmj.o622.
37. Swisher S. Florida surgeon general defends support of fringe group that touted false COVID "cure." Seattle Times, February 8, 2022, https://www.seattletimes.com/nation-world/florida-surgeon-general-defends-support-of-fringe-group-that-touted-false-covid-cure.
38. America's Frontline Doctors. Homepage, https://americasfrontlinedoctors.org; https://americasfrontlinedoctors.org/about-us; mission statement, https://americasfrontlinedoctors.org/2/about-us/mission-statement and https://americasfrontlinedoctors.org/index/about-us/mission-statement.
39. Healy J, McCarthy L. "See you in court": G.O.P. governors express outrage and vow to fight Biden's vaccine requirements. New York Times, September 10, 2021, https://www.nytimes.com/2021/09/10/us/republican-governors-mandate-reaction.html.
40. Wermund B. "Not on my watch": Texas Republicans buck Biden's door-to-door vaccine drive. Houston Chronicle, July 7, 2021, https://www.houstonchronicle.com/politics/texas/article/Not-on-my-watch-Texas-Republicans-buck-16299133.php.
41. Mello MM, Gould IV WB, Duff B, Driscoll S. A look at the Supreme Court ruling on vaccination mandates. SLS Blogs, January 20, 2022, https://law.stanford.edu/2022/01/20/a-look-at-the-supreme-court-ruling-on-vaccination-mandates.

Literature Cited

42. Kamarck E. Pandemic politics: Red state governors are in trouble for their Covid leadership. Brookings Institution, October 19, 2021, https://www.brookings.edu/blog/fixgov/2021/10/19/pandemic-politics-red-state-governors-are-in-trouble-for-their-covid-leadership.
43. Treisman R. What to know about Judge Kathryn Mizelle, who struck down the travel mask mandate. National Public Radio, April 19, 2022, https://www.npr.org/2022/04/19/1093566982/florida-mask-mandate-judge-kathryn-mizelle.
44. United States Department of Justice, Office of Public Affairs. Justice Department issues statement on ruling in Health Freedom Defense Fund Inc. et al. v. Biden et al., April 19, 2022, https://www.justice.gov/opa/pr/justice-department-issues-statement-ruling-health-freedom-defense-fund-inc-et-al-v-biden-et.
45. Health Freedom Defense Fund. HFDF lawsuits, https://healthfreedomdefense.org/hfdf-lawsuits.
46. Mueller J. Fauci, Biden official served subpoenas in lawsuit over collusion to suppress free speech. Center Square, July 20, 2022, https://www.thecentersquare.com/missouri/fauci-biden-officials-served-subpoenas-in-lawsuit-over-collusion-to-suppress-free-speech/article_29d5a5a0-07ab-11ed-afee-ef19381f2671.html.
47. Ellefson L. 60% of Fox News' vaccine reports this summer included anti-vax claims, study finds. Wrap, August 20, 2021, https://www.thewrap.com/media-matters-fox-news-vaccine-study.
48. Holmes J. Fox News is moving from "just asking questions" to full-on anti-vax crapola. Esquire, July 8, 2021, https://www.esquire.com/news-politics/a36967699/fox-news-anti-vax-tucker-carlson-just-asking-questions.
49. Savillo R, Monroe T. Fox's efforts to undermine vaccines has only worsened. Media Matters for America, August 19, 2021, https://www.mediamatters.org/fox-news/foxs-effort-undermine-vaccines-has-only-worsened.
50. Amore S. Tucker Compares COVID Vaccines to "Sterilization or Frontal Lobotomies" (video). The Wrap, July 27, 2021, https://www.thewrap.com/tucker-carlson-compares-covid-vaccines-to-sterilization-or-frontal-lobotomies.
51. Darcy O. Fox has quietly implemented its own version of a vaccine passport while its top personalities attack them. CNN Business, July 20, 2021, https://www.cnn.com/2021/07/19/media/fox-vaccine-passport/index.html.

52. Steinberg J. An all-out attack on "conservative misinformation." New York Times, October 31, 2008, https://www.nytimes.com/2008/11/01/washington/01media.html.
53. Blake A. Fox News's bad vaccine tweet, and what it portends. Washington Post, February 7, 2022, https://www.washingtonpost.com/politics/2022/02/07/fox-newss-bad-vaccine-tweet-what-it-portends.
54. Pinna M, Picard L, Goessmann C. Cable news and COVID-19 vaccine uptake. Sci Rep 12(1) (October 7, 2022): 16804, doi: 10.1038/s41598-022-20350-0.
55. Bump P. How much of right-wing opposition to vaccination was Fox News's fault? Washington Post, October 10, 2022, https://www.washingtonpost.com/politics/2022/10/10/vaccines-coronavirus-fox-news.
56. Confessore N. How Tucker Carlson stoked white fear to conquer cable. New York Times, April 30, 2022, https://www.nytimes.com/2022/04/30/us/tucker-carlson-gop-republican-party.html.
57. Kleefeld E. Fox News boss Lachlan Murdoch supports Tucker Carlson's misinformation against the COVID-19 vaccines. Media Matters, May 19, 2021, https://www.mediamatters.org/lachlan-murdoch/fox-news-boss-lachlan-murdoch-supports-tucker-carlsons-misinformation-against-covid.
58. Katz AJ. Top cable news shows of 2021: Tucker Carlson Tonight is no. 1 in all measurements for first time ever. AdWeek TV Newser, January 3, 2022, https://www.adweek.com/tvnewser/top-cable-news-shows-of-2021-tucker-carlson-tonight-is-no-1-in-all-categories-for-first-time-ever/496940.
59. Ladapo JA, Risch HA. Are Covid vaccines riskier than advertised? Wall Street Journal, June 22, 2021, https://www.wsj.com/articles/are-covid-vaccines-riskier-than-advertised-11624381749.
60. Bendavid E, Bhattacharya J. Is the coronavirus as deadly as they say? Wall Street Journal, May 24, 2020, https://www.wsj.com/articles/is-the-coronavirus-as-deadly-as-they-say-11585088464.
61. Makary M. We'll have herd immunity by April. Wall Street Journal, February 18, 2021, https://www.wsj.com/articles/well-have-herd-immunity-by-april-11613669731.
62. Knott M. Former Murdoch exec slams Fox News over vaccine misinformation. Sydney Morning Herald, July 18, 2021, https://www.smh.com.au/world/north-america/former-fox-news-exec-slams-network-over-vaccine-misinformation-20210717-p58ain.html.

63. Dowd M. James Murdoch, rebellious scion. New York Times, October 10, 2020, https://www.nytimes.com/2020/10/10/style/james-murdoch-maureen-dowd.html.
64. Li K. Murdoch receives COVID-19 vaccine as Fox News host casts suspicion on campaign. Reuters, December 18, 2020, https://www.reuters.com/article/us-health-coronavirus-murdoch/murdoch-receives-covid-19-vaccine-as-fox-news-host-casts-suspicion-on-campaign-idUSKBN28S2J7.
65. D'Ambrosio A. New institute has ties to the Great Barrington Declaration. MedPage Today, November 12, 2021, https://www.medpagetoday.com/special-reports/exclusives/95601.
66. McCool A, Wepukhulu KS. US conservatives spreading anti-vax misinformation to unvaccinated Uganda. Open Democracy, January 21, 2022, https://www.opendemocracy.net/en/5050/us-conservatives-spread-anti-vaccine-covid-misinformation-uganda.
67. Orac. The Brownstone Institute: Promoting antivaccine misinformation in Africa. Respectful Insolence (blog), January 31, 2022, https://respectfulinsolence.com/2022/01/31/brownstone-institute-promoting-antivaccine-misinformation-in-africa.
68. Yamey G, Gorski DH. Covid-19 and the new merchants of doubt. BMJ Opinion, September 13, 2021, https://blogs.bmj.com/bmj/2021/09/13/covid-19-and-the-new-merchants-of-doubt.
69. Bragman W, Kotch A. How the Koch network hijacked the war on COVID. The Lever, December 22, 2021, https://www.levernews.com/how-the-koch-network-hijacked-the-war-on-covid.
70. Orac. Jeffrey Tucker: The antivaccine movement and the far right. Respectful Insolence (blog), August 3, 2022, https://www.respectfulinsolence.com/2022/08/03/jeffrey-tucker-the-antivaccine-movement-and-the-far-right.
71. Gertz M. Alex Berenson's antivax commentary is killing Fox viewers because that's what the network brass wants. Media Matters, January 27, 2022, https://www.mediamatters.org/tucker-carlson/alex-berensons-antivax-commentary-killing-fox-viewers-because-thats-what-network.
72. O'Connell O. Outrage as Alex Berenson baselessly tells Tucker Carlson "dangerous" vaccines should be "withdrawn." Yahoo and Independent, January 26, 2022, https://www.yahoo.com/video/outrage-alex-berenson-baselessly-tells-163910649.html.
73. Tiffany K. A prominent vaccine skeptic returns to Twitter. Atlantic, Au-

gust 24, 2022, https://www.theatlantic.com/technology/archive/2022/08/alex-berenson-twitter-ban-lawsuit-covid-misinformation/671219.
74. Thompson D. The pandemic's wrongest man. Atlantic, April 1, 2021, https://www.theatlantic.com/ideas/archive/2021/04/pandemics-wrongest-man/618475.
75. Dickson EJ. "A menace to public health": Doctors demand Spotify puts [sic] an end to Covid lies on "Joe Rogan Experience." Rolling Stone, January 12, 2022, https://www.rollingstone.com/culture/culture-news/covid-misinformation-joe-rogan-spotify-petition-1282240.
76. Brumfiel G. Inside the growing alliance between anti-vaccine activists and pro-Trump Republicans. NPR, All Things Considered, December 6, 2021, https://www.npr.org/2021/12/06/1057344561/anti-vaccine-activists-political-conference-trump-republicans.
77. Bergengruen V. How the anti-vax movement is taking over the right. Time, January 26, 2022, https://time.com/6141699/anti-vaccine-mandate-movement-rally.
78. Orso A, Roebuck J. Feds charge an N.J. anti-vax activist and a correctional officer in Capitol riot. Philadelphia Inquirer, January 22, 2021, https://www.inquirer.com/news/new-jersey/stephanie-hazelton-capitol-insurrection-medford-new-jersey-rioters-20210122.html.
79. Reilly RJ. Anti-vaccine doctor who pushed hydroxychloroquine pleads guilty for entering Capitol on Jan. 6. NBC News, March 3, 2022, https://www.nbcnews.com/politics/justice-department/anti-vaccine-doctor-pushed-hydroxychloroquine-pleads-guilty-entering-c-rcna18535.
80. Jacobs S. Marine charged in Jan. 6 riot is arrested in N.Y. for selling forged vaccine cards to unvaccinated, including other military members. Washington Post, February 17, 2022, https://www.washingtonpost.com/national-security/2022/02/17/marine-charged-jan-6-riot-arrested-ny-selling-forged-vaccine-cards-unvaccinated-including-other-military-members.
81. Myers SL, Sullivan E. Disinformation has become another untouchable problem in Washington. New York Times, July 6, 2022, https://www.nytimes.com/2022/07/06/business/disinformation-board-dc.html.

5 | A Tough Time to Be a Scientist

1. Hotez PJ. The global fight to develop antipoverty vaccines in the antivaccine era. Hum Vaccin Immunother 14(9) (2018): 2128–31, doi: 10.1080/21645515.2018.1430542.

Literature Cited

2. Hotez PJ, Bottazzi ME. A COVID vaccine for all. Sci Am, December 30, 2021, https://www.scientificamerican.com/article/a-covid-vaccine-for-all.
3. Gray L. Dr. Peter Hotez's battle against the "anti-science confederacy" is a lifetime in the making. Houston Chronicle, February 27, 2021, https://www.houstonchronicle.com/life/article/Peter-Hotez-fight-for-science-COVID-houston-texas-15982489.php.
4. Facebook, https://www.facebook.com/resultswithmary/videos/10163288539190142, accessed May 1, 2022.
5. Hotez PJ. Texas and its measles epidemics. PLOS Med 13(10) (October 25, 2016): e1002153, doi: 10.1371/journal.pmed.1002153.
6. Biel L. Peter Hotez vs measles and the antivaccination movement. Texas Monthly, December 2017, https://www.texasmonthly.com/news-politics/scientist-stop-measles-texas.
7. Stickland J. Twitter, May 7, 2019, https://twitter.com/RepStickland/status/1125790483895279617, accessed May 15, 2022.
8. Stickland J. Twitter, May 7, 2019, https://twitter.com/RepStickland/status/1125830765961457664, accessed May 15, 2022.
9. Paul D. GOP state legislator attacks vaccine scientist on Twitter, accusing him of self-enrichment, "sorcery." Washington Post, May 8, 2019, https://www.washingtonpost.com/health/2019/05/08/gop-legislator-attacks-top-vaccine-scientist-twitter-accusing-him-self-enrichment-sorcery.
10. Bonifield J. Texas lawmaker calls vaccine research "sorcery." CNN, May 9, 2019, https://www.cnn.com/2019/05/09/health/texas-vaccines-sorcery-twitter/index.html.
11. Interlandi J. When defending vaccines gets ugly. New York Times, June 2, 2019, https://www.nytimes.com/2019/06/02/opinion/vaccines-peter-hotez.html.
12. Hotez P. COVID vaccines: Time to confront anti-vax aggression. Nature 592(7856) (April 2021): 661, doi: 10.1038/d41586-021-01084-x.
13. Gray L. Online trolls take anti-vaxx hate speech to a new level, attacking Houston's Dr. Peter Hotez. Houston Chronicle, May 7, 2021, https://www.houstonchronicle.com/news/houston-texas/health/article/Trolls-take-anti-vaxx-hate-speech-to-a-new-level-16160138.php.
14. Adams M. NATURE publishes insane rant by Texas pediatrician Peter Hotez, who seemingly calls for United Nations SHOCK TROOPS to wage "counteroffensive" against all anti-vaxxers—(opinion). Natural News, May 6, 2021, https://www.naturalnews.com/2021-05-06-nature-publishes-insane-rant-by-texas-pediatrician-peter-hotez.html.

Literature Cited

15. Hotez PJ. Vaccinating Cassandra. EClinicalMedicine 31 (January 2021): 100711, doi: 10.1016/j.eclinm.2020.100711.
16. Hotez PJ. Global vaccinations: New urgency to surmount a triple threat of illness, antiscience, and anti-Semitism. Rambam Maimonides Med J 14(1) (January 29, 2023): e0004, doi: 10.5041/RMMJ.10491.
17. Podhoretz J. It's October, so there's anti-semitism. Commentary (podcast), October 18, 2022, https://www.commentary.org/john-podhoretz/its-october-so-theres-anti-semitism.
18. Kampeas R. Fox Nation host compares Fauci to Nazi doctor Mengele. Jerusalem Post, December 1, 2021, https://www.jpost.com/diaspora/antisemitism/fox-news-host-compares-fauci-to-nazi-doctor-mengele-687484.
19. Gajewski R. Lara Logan dropped by UTA after comparing Anthony Fauci to infamous Nazi doctor. Hollywood Reporter, January 17, 2022, https://www.hollywoodreporter.com/business/business-news/lara-logan-dropped-uta-fauci-fox-news-1235076400.
20. Auschwitz Museum, Twitter, November 30, 2021, https://twitter.com/AuschwitzMuseum/status/1465549524542410753, accessed May 15, 2022.
21. Grynbaum MM. Fox News host's incendiary Fauci comments follow a network pattern. New York Times, December 23, 2021, https://www.nytimes.com/2021/12/23/business/media/fox-anthony-fauci-jesse-watters.html.
22. Treisman R. Fauci calls on Fox News to fire host who urged "kill shot" in "ambush" interviews. NPR Houston Public Media, December 22, 2021, https://www.npr.org/2021/12/22/1066956407/anthony-fauci-jesse-watters-fox-news-kill-shot-ambush-interviews.
23. Bella T. Tucker Carlson falsely claims Anthony S. Fauci "created" Covid. Washington Post, July 29, 2021, https://www.washingtonpost.com/health/2021/07/29/tucker-carlson-fauci-created-covid.
24. Cohen J. Almost everything Tucker Carlson said about Anthony Fauci this week was misleading or false. Science, August 25, 2022, https://www.science.org/content/article/almost-everything-tucker-carlson-said-about-anthony-fauci-week-was-misleading-or-false.
25. Stiles A. Fact check: Ron DeSantis says "someone needs to grab that little elf" Fauci and "chuck him across the Potomac." Washington Free Beacon, August 25, 2022, https://freebeacon.com/satire/ron-desantis-fauci-elf-chuck.

Literature Cited

26. Naylor B. Dr. Fauci says GOP Sen. Paul's false accusations have sparked death threats. National Public Radio, January 11, 2022, https://www.npr.org/2022/01/11/1072110378/dr-fauci-says-gop-sen-pauls-false-accusations-have-sparked-death-threats.
27. Solis N. Merced man arrested on way to White House with rifle, "hit list" compiled from TikTok. Los Angeles Times, December 29, 2021, https://www.latimes.com/california/story/2021-12-29/merced-man-arrested-on-way-to-white-house-with-rifle-hit-list-compiled-from-tiktok.
28. Dress B. Republican Oklahoma candidate says "we should try Anthony Fauci and put him in front of a firing squad." The Hill, April 27, 2022, https://thehill.com/news/campaign/3468172-republican-oklahoma-candidate-says-we-should-try-anthony-fauci-and-put-him-in-front-of-a-firing-squad.
29. Patel V. Man who threatened to kill Fauci is sentenced to 3 years in prison. New York Times, August 4, 2022, https://www.nytimes.com/2022/08/04/us/fauci-thomas-connally-sentenced.html.
30. Baxter M. GITMO double-header execution: Anthony Fauci and Loretta Lynch. Real Raw News, April 25, 2022, https://realrawnews.com/2022/04/gitmo-double-header-execution-anthony-fauci-loretta-lynch.
31. Hotez PJ. The unique terror of being a COVID scientist after Jan. 6. Daily Beast, June 21, 2022, https://www.thedailybeast.com/the-unique-terror-of-being-a-covid-scientist-after-january-6.
32. Eban K. "This shouldn't happen": Inside the virus-hunting nonprofit at the center of the lab-leak controversy. Vanity Fair, March 31, 2022, https://www.vanityfair.com/news/2022/03/the-virus-hunting-nonprofit-at-the-center-of-the-lab-leak-controversy.
33. Maxmen A. Wuhan market was epicentre of pandemic's start, studies suggest. Nature 603 (February 27, 2022): 15–16, doi: https://doi.org/10.1038/d41586-022-00584-8.
34. Worobey M, Levy JI, Serrano LM, Crits-Christoph A, Pekar JE, Goldstein SA, Rasmussen AL, Kraemer MUG, Newman C, Koopmans MPG, Suchard MA, Wertheim JO, Lemey P, Robertson DL, Garry RF, Holmes EC, Rambaut A, Andersen KG. The Huanan seafood wholesale market in Wuhan was the early epicenter of the COVID-19 pandemic. Science 377(6609) (July 26, 2022): abp8715, doi: 10.1126/science.abp8715.
35. Mallapaty S. Coronaviruses closely related to the pandemic virus discovered in Japan and Cambodia. Nature 588 (2020): 15–16, doi: https://doi.org/10.1038/d41586-020-03217-0.

36. Liu X, Wu Q, Zhang Z. Global diversification and distribution of coronaviruses with furin cleavage sites. Front Microbiol 12 (October 7, 2021): 649314, doi: 10.3389/fmicb.2021.649314.
37. Wu Y, Zhao S. Furin cleavage sites naturally occur in coronaviruses. Stem Cell Res 50 (December 9, 2020): 102115, doi: 10.1016/j.scr.2020.102115.
38. Garry RF. The evidence remains clear: SARS-CoV-2 emerged via the wildlife trade. Proc Natl Acad Sci U S A 119(47) (November 22, 2022): e2214427119, doi: 10.1073/pnas.2214427119.
39. Garry RF. SARS-CoV-2 furin cleavage site was not engineered. Proc Natl Acad Sci U S A 119(40) (October 4, 2022): e2211107119, doi: 10.1073/pnas.2211107119.
40. Rasmussen A, Worobey M. Conspiracy theories about COVID-19 help nobody. Foreign Policy, September 15, 2022, https://foreignpolicy.com/2022/09/15/conspiracy-theories-covid-19-commission.
41. Keusch GT, Amuasi JH, Anderson DE, Daszak P, Eckerle I, Field H, Koopmans M, Lam SK, Das Neves CG, Peiris M, Perlman S, Wacharapluesadee S, Yadana S, Saif L. Pandemic origins and a One Health approach to preparedness and prevention: Solutions based on SARS-CoV-2 and other RNA viruses. Proc Natl Acad Sci U S A 119(42) (October 18, 2022): e2202871119, doi: 10.1073/pnas.2202871119.
42. Nives A. Progress! Congress defunds Wuhan animal lab. EcoHealth Alliance, White Coat Waste Project, June 24, 2022, https://blog.whitecoatwaste.org/2022/06/24/progress-congress-defunds-wuhan-animal-lab-ecohealth-alliance.
43. Weixel N. GOP plots Fauci probe after midterms. The Hill, July 25, 2022, https://thehill.com/policy/healthcare/3571232-gop-plots-fauci-probe-after-midterms.
44. Culliton BJ. Dingell v. Baltimore. Science 244(4903) (April 28, 1989): 412–14, doi: 10.1126/science.2655079.
45. New York Times News Service. Exoneration of scientist casts doubt on fraud agency: Decadelong case pitted Nobel laureate against powerful congressman. Baltimore Sun, June 24, 1996, https://www.baltimoresun.com/news/bs-xpm-1996-06-25-1996177012-story.html.
46. Lantos PM, Charini WA, Medoff G, Moro MH, Mushatt DM, Parsonnet J, Sanders JW, Baker CJ. Final report of the Lyme disease review panel of the Infectious Diseases Society of America. Clin Infect Dis 51(1) (July 1, 2010): 1–5, doi: 10.1086/654809.
47. Paul R. #FireFauci, https://randpaul.com/issue/firefauci.

Literature Cited

48. Newburger E. Dr. Fauci says his daughters need security as family continues to get death threats. CNBC, August 5, 2020, https://www.cnbc.com/2020/08/05/dr-fauci-says-his-daughters-need-security-as-family-continues-to-get-death-threats.html.
49. Nogrady B. "I hope you die": How the COVID pandemic unleashed attacks on scientists. Nature 598 (2021): 250–53, doi: https://doi.org/10.1038/d41586-021-02741-x.
50. O'Grady C. In the line of fire. Science 375(6587) (March 24, 2022): https://www.science.org/content/article/overwhelmed-hate-covid-19-scientists-face-avalanche-abuse-survey-shows.
51. Fox News. Transcript: Ingraham on rising crime in US cities, critical race theory. Fox News, June 23, 2021, https://www.foxnews.com/transcript/ingraham-on-rising-crime-in-us-cities-critical-race-theory.
52. Pennyfarthing AJ. Ron DeSantis fed COVID crow by doctor he'd ridiculed on Fox News, Daily Kos, July 15, 2021, https://www.dailykos.com/stories/2021/7/15/2040161/-Ron-DeSantis-fed-COVID-crow-by-doctor-he-d-ridiculed-on-Fox-News.
53. Edwards D. Marjorie Taylor Greene: "Scientists have been wrong over and over and over since the beginning of time." Raw Story, January 13, 2022, https://www.rawstory.com/marjorie-taylor-greene-scientists.
54. Ibsen H. An Enemy of the People. Mineola, NY: Dover Thrift Editions, 1999.
55. Hotez P. Who will defend embattled scientists? Boston Globe, January 27, 2022, https://www.bostonglobe.com/2022/01/27/opinion/who-will-defend-embattled-scientists.
56. Creitz C. Tucker Carlson slams COVID "nutcase from Baylor" Peter Hotez for "discrediting American medicine." Fox News, February 1, 2022, https://www.foxnews.com/media/tucker-carlson-covid-peter-hotez.
57. Fox News. Transcript: "The Ingraham Angle" on Jeff Zucker. Fox News, February 4, 2022, https://www.foxnews.com/transcript/ingraham-angle-zucker.
58. American Heritage Foundation. HUAC and the Manhattan Project, July 15, 2016, https://www.atomicheritage.org/history/huac-and-manhattan-project.
59. Wang J. American Science in an Age of Anxiety: Scientists, Anticommunism, and the Cold War. Chapel Hill: University of North Carolina Press, 1999.

60. Hotez PJ. Mounting antiscience aggression in the United States. PLOS Biol 19(7) (2021): e3001369, https://doi.org/10.1371/journal.pbio.3001369.
61. Diaz D, Grayer A, Nobles R, LeBlanc P. Marjorie Taylor Greene launching "America First" caucus pushing for "Anglo-Saxon political tradition." CNN, April 17, 2021, https://www.cnn.com/2021/04/16/politics/marjorie-taylor-greene-america-first-caucus/index.html.

6 | The Authoritarian Playbook

1. Baragona J. GOP senator blasts Tucker Carlson's pro-Russia rhetoric: "We side" with democracies. Daily Beast, January 30, 2022, https://www.thedailybeast.com/gop-sen-james-risch-criticizes-tucker-carlsons-pro-russia-rhetoric-says-we-side-with-democracies.
2. MSNBC. How Tucker Carlson became one of Russia's biggest cheerleaders. All in with Chris Hayes, February 21, 2022, https://www.msnbc.com/all-in/watch/how-tucker-carlson-became-one-of-russia-s-biggest-cheerleaders-133713989787.
3. Pilkington E. Tucker Carlson viewers calling me to say US should back Russia, Democrat says. Guardian, January 25, 2022, https://www.theguardian.com/world/2022/jan/25/tucker-carlson-russia-ukraine-democrats.
4. Saletan W. Father Carlson: Tucker Carlson is on Russia's side. Bulwark, February 23, 2022, https://www.thebulwark.com/father-carlson.
5. Thompson SA. How Russian media uses Fox News to make its case. New York Times, April 15, 2022, https://www.nytimes.com/2022/04/15/technology/russia-media-fox-news.html.
6. Graziosi G. Tucker Carlson hits back at claims he's a Russia propagandist—despite previously saying he backed Putin. Independent, March 15, 2022, https://www.independent.co.uk/news/world/americas/us-politics/tucker-carlson-fox-news-putin-russia-b2036335.html.
7. Mastrangelo D. Panel on "The View" calls for DOJ to probe Tucker Carlson over Putin rhetoric. The Hill, March 14, 2022, https://thehill.com/homenews/media/598113-panel-on-the-view-call-for-doj-to-probe-tucker-carlson-over-putin-rhetoric.
8. Swearingen K. Won't someone, anyone stand up to protest Tucker Carlson, Putin's biggest fanboy? Salon, March 21, 2022, https://www.salon.com/2022/03/21/wont-someone-anyone-stand-up-to-tucker-carlson-putins-biggest-fanboy.

9. Ings S. Stalin and the Scientists: A History of Triumph and Tragedy 1905–1953. New York: Grove Press, 2016.
10. Kaempffert W. Science in the totalitarian state. Foreign Affairs 19(2) (1941): 433–41.
11. Arendt H. The Origins of Totalitarianism. New York: Schocken Books, 1951.
12. Pringle P. The Murder of Nikolai Vavilov. New York: Simon and Schuster, 2008.
13. Hotez PJ. Anti-science kills: From Soviet embrace of pseudoscience to accelerated attacks on US biomedicine. PLOS Biol 19 (2021): e3001068.
14. Lourie R. Sakharov, a Biography. Waltham, MA: Brandeis University Press, 2002.
15. Catanzaro M. Alarm as execution looms for scientist on death row in Iran. Nature, November 30, 2020, https://www.nature.com/articles/d41586-020-03396-w.
16. Applebaum A. Twilight of Democracy: The Seductive Lure of Authoritarianism. New York: Doubleday, 2020.
17. Santora M. George Soros-founded university is forced out of Hungary. New York Times, December 3, 2018, https://www.nytimes.com/2018/12/03/world/europe/soros-hungary-central-european-university.html.
18. Serdult V. Tucker Carlson has become obsessed with Hungary: Here's what he doesn't understand. Politico, February 2, 2022, https://www.politico.com/news/magazine/2022/02/01/tucker-carlson-hungary-orban-00004149.
19. Marantz A. Does Hungary offer a glimpse of our authoritarian future? New Yorker, July 4, 2022, https://www.newyorker.com/magazine/2022/07/04/does-hungary-offer-a-glimpse-of-our-authoritarian-future.
20. Beauchamp Z. Viktor Orbán laid out his dark worldview to the American right—and they loved it. Vox, August 5, 2022, https://www.vox.com/2022/8/5/23292448/orban-cpac-dallas-2022-speech-trump.
21. Schepple quoted in: Eisler P, Ulmer A, Komuves A, Marshall ARc. U.S. conservative conference with Hungary's hardline leader reflects Republican divide. Reuters, April 5, 2022, https://www.reuters.com/world/us/us-conservative-conference-with-hungarys-hardline-leader-reflects-republican-2022-04-05.
22. Kovensky J. What is CPAC doing in Brazil? TMP (Talking Points Memo), September 7, 2021, https://talkingpointsmemo.com/muckraker/what-is-cpac-doing-in-brazil.

23. Taylor L. "We are being ignored": Brazil's researchers blame anti-science government for devastating COVID surge. Nature 593 (April 27, 2021): 15–16, https://www.nature.com/articles/d41586-021-01031-w.
24. Barbara V. Telegram groups are wild, sinister places. New York Times, May 4, 2022, https://www.nytimes.com/2022/05/04/opinion/bolsonaro-brazil-telegram-misinformation.html.
25. Molla R. Why right-wing extremists' favorite new platform is so dangerous. Vox, January 20, 2021, https://www.vox.com/recode/22238755/telegram-messaging-social-media-extremists.
26. Ben-Ghiat R. Vaccine scientist Peter Hotez on anti-science aggression. Substack, November 17, 2021, https://lucid.substack.com/p/vaccine-scientist-peter-hotez-on?utm_source=url.
27. Ben-Ghiat R. Strongmen: Mussolini to the Present. New York: WW Norton, 2020.
28. Jong-Fast M. GOP's anti-vaxx disinfo "is what authoritarian regimes do" w/ Dr. Peter Hotez and James Carville. New Abnormal (podcast), https://podcasts.apple.com/us/podcast/gops-anti-vaxx-disinfo-is-what-authoritarian-regimes/id1508202790?i=1000530829694.
29. Lasco G. Medical populism and the COVID-19 pandemic. Glob Public Health 15(10) (October 2020): 1417–29, doi: https://doi.org/10.1080/17441692.2020.1807581.
30. McNicoll T. In vaccine-sceptic France, candidates walk tightrope on Covid measures. France 24, February 16, 2022, https://www.france24.com/en/europe/20220216-in-vaccine-sceptic-france-candidates-walk-a-tightrope-on-covid-measures.
31. Hotez P. COVID vaccines: Time to confront anti-vax aggression. Nature 592(7856) (April 2021): 661, doi: 10.1038/d41586-021-01084-x.
32. Gordon MR, Volz D. Russian disinformation campaign aims to undermine confidence in Pfizer, other Covid-19 vaccines, U.S. officials say. Wall Street Journal, March 7, 2021, https://www.wsj.com/articles/russian-disinformation-campaign-aims-to-undermine-confidence-in-pfizer-other-covid-19-vaccines-u-s-officials-say-11615129200.
33. Grimes DR. Russian misinformation seeks to confound, not convince. Sci Am, March 28, 2022, https://www.scientificamerican.com/article/russian-misinformation-seeks-to-confound-not-convince.
34. Paul C, Matthews M. The Russian "firehose of falsehood" propaganda model: Why it might work and options to counter it. RAND Corporation, 2016, https://www.rand.org/pubs/perspectives/PE198.html.

35. Rankin J. EU says China behind "huge wave" of Covid-19 disinformation. Guardian, June 10, 2020, https://www.theguardian.com/world/2020/jun/10/eu-says-china-behind-huge-wave-covid-19-disinformation-campaign.
36. US Virtual Embassy Iran. Iran: COVID-19 disinformation fact sheet, https://ir.usembassy.gov/iran-covid-19-disinformation-fact-sheet.
37. Broniatowski DA, Jamison AM, Qi S, AlKulaib L, Chen T, Benton A, Quinn SC, Dredze M. Weaponized health communication: Twitter bots and Russian trolls amplify the vaccine debate. Am J Public Health 108(10) (October 2018): 1378–84, doi: 10.2105/AJPH.2018.304567.
38. Broad WJ. Putin's long war against American science. New York Times, April 13, 2020, updated June 16, 2021, https://www.nytimes.com/2020/04/13/science/putin-russia-disinformation-health-coronavirus.html
39. Butler K. How wellness influencers became cheerleaders for Putin's war. Mother Jones, March 24, 2022, https://www.motherjones.com/politics/2022/03/how-wellness-influencers-became-cheerleaders-for-putins-war.
40. Dettmer J. Russian anti-vaccine disinformation campaign backfires. VOA News, November 18, 2021, https://www.voanews.com/a/russian-anti-vaccine-disinformation-campaign-backfires/6318536.html.
41. Troijanovski A. "You can't trust anyone": Russia's hidden Covid toll is an open secret. New York Times, April 10, 2021, updated October 18, 2021, https://www.nytimes.com/2021/04/10/world/europe/covid-russia-death.html.
42. COVID-19 Excess Mortality Collaborators. Estimating excess mortality due to the COVID-19 pandemic: A systematic analysis of COVID-19-related mortality, 2020–21. Lancet 399(10334) (April 16, 2022): 1513–36, doi: 10.1016/S0140-6736(21)02796-3.
43. Hotez PJ. Russian–United States vaccine science diplomacy: Preserving the legacy. PLOS Negl Trop Dis 11(5) (2017): e0005320, https://doi.org/10.1371/journal.pntd.0005320.
44. Devega C. Global forecaster on "another bad year for democracy": Is the world near a dire tipping point? Salon, May 2, 2022, https://www.salon.com/2022/05/02/global-forecaster-on-another-year-for-democracy-is-the-world-near-a-dire-tipping-point.

7 | The Hardest Science Communication Ever

1. Requarth T. Pandemic orphans are slipping through the cracks. Atlantic, April 6, 2022, https://www.theatlantic.com/health/archive/2022/04/covid-orphan-kids-lost-parent/629436.
2. Nalbandian A, Sehgal K, Gupta A, Madhavan MV, McGroder C, Stevens JS, Cook JR, Nordvig AS, Shalev D, Sehrawat TS, Ahluwalia N, Bikdeli B, Dietz D, Der-Nigoghossian C, Liyanage-Don N, Rosner GF, Bernstein EJ, Mohan S, Beckley AA, Seres DS, Choueiri TK, Uriel N, Ausiello JC, Accili D, Freedberg DE, Baldwin M, Schwartz A, Brodie D, Garcia CK, Elkind MSV, Connors JM, Bilezikian JP, Landry DW, Wan EY. Post-acute COVID-19 syndrome. Nat Med 27(4) (April 27, 2021): 601–15, doi: 10.1038/s41591-021-01283-z.
3. Douaud G, Lee S, Alfaro-Almagro F, Arthofer C, Wang C, McCarthy P, Lange F, Andersson JLR, Griffanti L, Duff E, Jbabdi S, Taschler B, Keating P, Winkler AM, Collins R, Matthews PM, Allen N, Miller KL, Nichols TE, Smith SM. SARS-CoV-2 is associated with changes in brain structure in UK Biobank. Nature, March 7, 2022, doi: 10.1038/s41586-022-04569-5.
4. Antonelli M, Penfold RS, Merino J, Sudre CH, Molteni E, Berry S, Canas LS, Graham MS, Klaser K, Modat M, Murray B, Kerfoot E, Chen L, Deng J, Österdahl MF, Cheetham NJ, Drew DA, Nguyen LH, Pujol JC, Hu C, Selvachandran S, Polidori L, May A, Wolf J, Chan AT, Hammers A, Duncan EL, Spector TD, Ourselin S, Steves CJ. Risk factors and disease profile of post-vaccination SARS-CoV-2 infection in UK users of the COVID Symptom Study app: A prospective, community-based, nested, case-control study. Lancet Infect Dis 22(1) (September 1, 2021): P43–55, doi: 10.1016/S1473-3099(21)00460-6.
5. Hotez P, Batista C, Amor YB, Ergonul O, Figueroa JP, Gursel M, Hassanain M, Kang G, Kaslow DC, Kim JH, Lall B, Larson H, Sheahan T, Shoham S, Wilder-Smith A, Sow SO, Yadav P, Bottazzi ME. Should we vaccinate against long-COVID? Vaccine Insights 1(1) (2022): 101–6, doi: 10.18609/vac.2022.016.
6. Sheikh K. Covid depression is real: Here's what you need to know. New York Times, November 12, 2022, https://www.nytimes.com/2022/11/12/well/long-covid-depression-symptoms-treatment.html.
7. Frankovic K. Fewer than half of Republicans now support requiring childhood vaccinations for infectious diseases. YouGovAmerica, October 13,

Literature Cited

2021, https://today.yougov.com/topics/politics/articles-reports/2021/10/13/support-requiring-child-vaccinations.
8. Moore JP. Op-ed: The anti-vax movement was already getting scary: COVID supercharged it. Los Angeles Times, February 25, 2022, https://www.latimes.com/opinion/story/2022-02-25/covid-anti-vax-childhood-vaccination-measles-mumps.
9. National Public Radio and Associated Press. When Tennessee fired its vaccine chief, officials were caught off guard, emails show. NPR, October 26, 2021, https://www.npr.org/2021/10/26/1049321391/tennessee-vaccine-chief-fired-officials-emails.
10. Messerly M, Mahr K. Covid vaccine concerns are starting to spill over into routine immunizations. Politico, April 18, 2022, https://www.politico.com/news/2022/04/18/kids-are-behind-on-routine-immunizations-covid-vaccine-hesitancy-isnt-helping-0002550.
11. Hotez PJ, Nuzhath T, Colwell B. Combating vaccine hesitancy and other 21st century social determinants in the global fight against measles. Curr Opin Virol 41 (April 2020): 1–7, doi: 10.1016/j.coviro.2020.01.001.
12. Wilkinson E. Is anti-vaccine sentiment affecting routine childhood immunisations? BMJ 376 (February 10, 2022): e360, https://doi.org/10.1136/bmj.o360.
13. World Health Organization. Immunization coverage: Key facts, July 14, 2022, https://www.who.int/news-room/fact-sheets/detail/immunization-coverage.
14. Hotez PJ. Will anti-vaccine activism in the USA reverse global goals? Nat Rev Immunol 22 (August 1, 2022): 525–26, doi: 10.1038/s41577-022-00770-9.
15. Rigby J. Measles cases jump 79% in 2022 after COVID hit vaccination campaigns. Reuters, April 27, 2022, https://www.reuters.com/business/healthcare-pharmaceuticals/measles-cases-jump-79-2022-after-covid-hit-vaccination-campaigns-2022-04-27.
16. UN News. UN condemns brutal killing of eight polio workers in Afghanistan, February 24, 2022, https://news.un.org/en/story/2022/02/1112612.
17. Branswell H. Polioviruses found in wastewater samples in 2 N.Y. counties, suggesting continued spread. Stat News, August 4, 2022, https://www.statnews.com/2022/08/04/polioviruses-found-in-wastewater-samples-in-2-n-y-counties-suggesting-continued-spread.

18. Wise J. Poliovirus is detected in sewage from north and east London. BMJ 377 (June 23, 2022): o1546, doi: 10.1136/bmj.o1546.
19. Anthes E. Polio may have been spreading in New York since April. New York Times, August 16, 2022, https://www.nytimes.com/2022/08/16/health/polio-new-york.html.
20. Link-Gelles R, Lutterloh E, Ruppert PS, Backenson PB, St George K, Rosenberg ES, Anderson BJ, Fuschino M, Popowich M, Punjabi C, Souto M, McKay K, Rulli S, Insaf T, Hill D, Kumar J, Gelman I, Jorba J, Ng TFF, Gerloff N, Masters NB, Lopez A, Dooling K, Stokley S, Kidd S, Oberste MS, Routh J; 2022 U.S. Poliovirus Response Team. Public health response to a case of paralytic poliomyelitis in an unvaccinated person and detection of poliovirus in wastewater—New York, June–August 2022. Am J Transplant 22(10) (October 2022): 2470–74, doi: 10.1111/ajt.16677.
21. Ryerson AB, Lang D, Alazawi MA, Neyra M, Hill DT, St George K, Fuschino M, Lutterloh E, Backenson B, Rulli S, Ruppert PS, Lawler J, McGraw N, Knecht A, Gelman I, Zucker JR, Omoregie E, Kidd S, Sugerman DE, Jorba J, Gerloff N, Ng TFF, Lopez A, Masters NB, Leung J, Burns CC, Routh J, Bialek SR, Oberste MS, Rosenberg ES; 2022 U.S. Poliovirus Response Team. Wastewater testing and detection of poliovirus type 2 genetically linked to virus isolated from a paralytic polio case—New York, March 9–October 11, 2022. MMWR Morb Mortal Wkly Rep 71(44) (November 4, 2022): 1418–24, doi: 10.15585/mmwr.mm7144e2.
22. Malik AA, Winters MS, Omer S. Attitudes of the US general public towards monkeypox. Medr iv, June 20, 2022, https://www.medrxiv.org/content/10.1101/2022.06.20.22276527v1.
23. Carter J, Desikan A, Coldman G. The Trump administration has attacked science 100 times . . . and counting. Sci Am, May 29, 2019, https://blogs.scientificamerican.com/observations/the-trump-administration-has-attacked-science-100-times-and-counting.
24. Union of Concerned Scientists. Fetal tissue research blocked by a biased advisory committee, September 14, 2020, https://www.ucsusa.org/resources/attacks-on-science/fetal-tissue-research-blocked.
25. Hotez PJ. Combating antiscience: Are we preparing for the 2020s? PLOS Biol 18(3) (March 27, 2020): e3000683, doi: 10.1371/journal.pbio.3000683.
26. Scheufele DA, Krause NM, Freiling I, Brossard D. What we know about effective public engagement on CRISPR and beyond. Proc Natl Acad Sci U S A 118(22) (June 1, 2021): e2004835117, doi: 10.1073/pnas.2004835117.

27. Research!America. Survey: Most Americans cannot name a living scientist or a research institution, May 11, 2021, https://www.researchamerica.org/blog/survey-most-americans-cannot-name-a-living-scientist-or-a-research-institution.
28. Reyna VF. A scientific theory of gist communication and misinformation resistance, with implications for health, education, and policy. Proc Natl Acad Sci U S A 118(15) (April 13, 2021): e1912441117, doi: 10.1073/pnas.1912441117.
29. Caulfield T. Covid vaccine and mask conspiracies succeed when they appeal to identity and ideology. NBC News, December 8, 2020, https://www.nbcnews.com/think/opinion/covid-vaccine-mask-conspiracies-succeed-when-they-appeal-identity-ideology-ncna1251761.
30. Scheufele DA, Hoffman AJ, Neeley L, Reid CM. Misinformation about science in the public sphere. Proc Natl Acad Sci U S A 118(15) (April 13, 2021): e2104068118, doi: 10.1073/pnas.2104068118.
31. Hotez PJ. Vaccines Did Not Cause Rachel's Autism. Baltimore, MD: Johns Hopkins University Press, 2018.
32. Jamison KR. An Unquiet Mind. New York: Vintage Books, 1995.
33. O'Grady C. In the line of fire. Science 375(6587) (March 24, 2022): https://www.science.org/content/article/overwhelmed-hate-covid-19-scientists-face-avalanche-abuse-survey-shows.
34. Van Noorden R. Higher-profile COVID experts more likely to get online abuse. Nature, April 4, 2022, doi: https://doi.org/10.1038/d41586-022-00936-4.
35. Hotez P. Communicating science and protecting scientists in a time of political instability. Trends Mol Med 28(3) (March 2022): 173–75, doi: 10.1016/j.molmed.2022.01.001.
36. Hotez PJ. The medical biochemistry of poverty and neglect. Mol Med 20 (Suppl 1) (December 16, 2014): S31–36, doi: 10.2119/molmed.2014.00169.
37. Adegnika AA, de Vries SG, Zinsou FJ, Honkepehedji YJ, Dejon Agobé JC, Vodonou KG, Bikangui R, Bouyoukou Hounkpatin A, Bache EB, Massinga Loembe M, van Leeuwen R, Molemans M, Kremsner PG, Yazdanbakhsh M, Hotez PJ, Bottazzi ME, Li G, Bethony JM, Diemert DJ, Grobusch MP; HookVac Consortium. Safety and immunogenicity of co-administered hookworm vaccine candidates Na-GST-1 and Na-APR-1 in Gabonese adults: A randomised, controlled, double-blind, phase 1 dose-escalation trial. Lancet Infect Dis 21(2) (February 2021): 275–85, doi: 10.1016/S1473-3099(20)30288-7.

38. Hotez PJ, Bottazzi ME, Bethony J, Diemert DD. Advancing the development of a human schistosomiasis vaccine. Trends Parasitol 35(2) (February 2019): 104–8, doi: 10.1016/j.pt.2018.10.005.
39. Bronowski J. Science and Human Values. New York: Harper & Row, 1956.
40. US National Academies of Science, Engineering, and Medicine. The Integration of the Humanities and Arts with Sciences, Engineering, and Medicine in Higher Education: Branches from the Same Tree. Washington, DC: National Academies Press, 2018, https://doi.org/10.17226/24988, accessed November 20, 2022.
41. Aenile J.; mentored and edited by Wolf L. Communicating science: Researchers share their fears and tips. National Association of Science Writers, May 10, 2022, https://www.nasw.org/article/communicating-science-researchers-share-their-fears-and-tips.
42. Oreskes N, Conway EM. Merchants of Doubt: How a Handful of Scientists Obscured the Truth on Issues from Tobacco Smoke to Global Warming. London: Bloomsbury Press, 2010.
43. Singer JA. Against scientific gatekeeping: Science should be a profession, not a priesthood. Cato Institute, March 16, 2022, https://www.cato.org/commentary/against-scientific-gatekeeping#. The article also appeared in the May 2022 issue of Reason, https://reason.com/2022/04/03/against-scientific-gatekeeping.
44. Mayer J. The Kochs vs. Cato. New Yorker, March 1, 2012, https://www.newyorker.com/news/news-desk/the-kochs-vs-cato.
45. Martins-Filho PR, Ferreira LC, Heimfarth L, Araújo AAS, Quintans-Júnior LJ. Efficacy and safety of hydroxychloroquine as pre- and post-exposure prophylaxis and treatment of COVID-19: A systematic review and meta-analysis of blinded, placebo-controlled, randomized clinical trials. Lancet Reg Health Am 2 (October 2021): 100062, doi: 10.1016/j.lana.2021.100062.
46. Axfors C, Schmitt AM, Janiaud P, Van't Hooft J, Abd-Elsalam S, Abdo EF, Abella BS, Akram J, Amaravadi RK, Angus DC, Arabi YM, Azhar S, Baden LR, Baker AW, Belkhir L, Benfield T, Berrevoets MAH, Chen CP, Chen TC, Cheng SH, Cheng CY, Chung WS, Cohen YZ, Cowan LN, Dalgard O, de Almeida e Val FF, de Lacerda MVG, de Melo GC, Derde L, Dubee V, Elfakir A, Gordon AC, Hernandez-Cardenas CM, Hills T, Hoepelman AIM, Huang YW, Igau B, Jin R, Jurado-Camacho F, Khan KS, Kremsner PG, Kreuels B, Kuo CY, Le T, Lin YC, Lin WP, Lin TH, Lyngbakken MN, McArthur C, McVerry BJ, Meza-Meneses P, Monteiro WM, Morpeth SC, Mourad A, Mulligan MJ, Murthy S, Naggie S, Narayanasamy S, Nichol A,

Literature Cited

Novack LA, O'Brien SM, Okeke NL, Perez L, Perez-Padilla R, Perrin L, Remigio-Luna A, Rivera-Martinez NE, Rockhold FW, Rodriguez-Llamazares S, Rolfe R, Rosa R, Røsjø H, Sampaio VS, Seto TB, Shahzad M, Soliman S, Stout JE, Thirion-Romero I, Troxel AB, Tseng TY, Turner NA, Ulrich RJ, Walsh SR, Webb SA, Weehuizen JM, Velinova M, Wong HL, Wrenn R, Zampieri FG, Zhong W, Moher D, Goodman SN, Ioannidis JPA, Hemkens LG. Mortality outcomes with hydroxychloroquine and chloroquine in COVID-19 from an international collaborative meta-analysis of randomized trials. Nat Commun 12(1) (April 15, 2021): 2349, doi: 10.1038/s41467-021-22446-z.

47. Amani B, Khanijahani A, Amani B. Hydroxychloroquine plus standard of care compared with standard of care alone in COVID-19: A meta-analysis of randomized controlled trials. Sci Rep 11(1) (June 7, 2021): 11974, doi: 10.1038/s41598-021-91089-3.

48. Avezum Á, Oliveira GBF, Oliveira H, Lucchetta RC, Pereira VFA, Dabarian AL, D'O Vieira R, Silva DV, Kormann APM, Tognon AP, De Gasperi R, Hernandes ME, Feitosa ADM, Piscopo A, Souza AS, Miguel CH, Nogueira VO, Minelli C, Magalhães CC, Morejon KML, Bicudo LS, Souza GEC, Gomes MAM, Fo JJFR, Schwarzbold AV, Zilli A, Amazonas RB, Moreira FR, Alves LBO, Assis SRL, Neves PDMM, Matuoka JY, Boszczowski I, Catarino DGM, Veiga VC, Azevedo LCP, Rosa RG, Lopes RD, Cavalcanti AB, Berwanger O; COPE—COALITION COVID-19 Brazil V Investigators. Hydroxychloroquine versus placebo in the treatment of non-hospitalised patients with COVID-19 (COPE—Coalition V): A double-blind, multicentre, randomised, controlled trial. Lancet Reg Health Am 11 (July 2022): 100243, doi: 10.1016/j.lana.2022.100243.

49. Orac. A risible attack on the "priesthood" of "scientific gatekeeping." Respectful Insolence Blog, April 6, 2022, https://respectfulinsolence.com/2022/04/06/a-risible-attack-on-scientific-gatekeeping.

50. Sagan C. Broca's Brain: Reflections on the Romance of Science. New York: Random House, 1979.

51. Yamey G, Gorski DH. Covid-19 and the new merchants of doubt. BMJ Opinion, September 13, 2021, https://blogs.bmj.com/bmj/2021/09/13/covid-19-and-the-new-merchants-of-doubt.

52. Smith TC, Novella SP. HIV denial in the Internet era. PLOS Med 4(8) (August 21,2007): e256, 2007. https://doi.org/10.1371/journal.pmed.0040256.

8 | Southern Poverty Law Center for Scientists

1. Gladwell M. The Tipping Point: How Little Things Can Make a Big Difference. Boston: Little, Brown, 2000.
2. Brigham B. Florida bans 28 math textbooks in panic over critical race theory. Raw Story, April 15, 2022, https://www.rawstory.com/desantis-critical-race-theory.
3. Hotez PJ. The antiscience movement is escalating, going global and killing thousands. Sci Am, March 29, 2021, https://www.scientificamerican.com/article/the-antiscience-movement-is-escalating-going-global-and-killing-thousands.
4. Hotez PJ. Biden battles a triple-headed monster on vaccines. Daily Beast, July 19, 2021, https://www.thedailybeast.com/biden-battles-a-triple-headed-monster-on-vaccines.
5. Krugman P. Why did Republicans become so extreme? New York Times, June 27, 2022, https://www.nytimes.com/2022/06/27/opinion/republicans-extreme-abortion.html.
6. Dias E, Graham R. The growing religious fervor in the American right: "This is a Jesus Movement." New York Times, April 6, 2022, updated April 11, 2022, https://www.nytimes.com/2022/04/06/us/christian-right-wing-politics.html.
7. Hochman N. What comes after the religious right? New York Times, June 1, 2022, https://www.nytimes.com/2022/06/01/opinion/republicans-religion-conservatism.html.
8. Adler-Bell S. The violent fantasies of Blake Master. New York Times, August 3, 2022, https://www.nytimes.com/2022/08/03/opinion/blake-masters-arizona-senate.html.
9. Adler-Bell S. The radical young intellectuals who want to take over the American right. New Republic, December 2, 2021, https://newrepublic.com/article/164408/young-intellectuals-illiberal-revolution-conservatism.
10. Callaghan T, Washburn D, Goidel K, Nuzhath T, Spiegelman A, Scobee J, Moghtaderi A, Motta M. Imperfect messengers? An analysis of vaccine confidence among primary care physicians. Vaccine 40(18) (April 20, 2002): 2588–603, doi: 10.1016/j.vaccine.2022.03.025.
11. DiResta R. The supply of disinformation will soon be infinite. Atlantic, September 20, 2020, https://www.theatlantic.com/ideas/archive/2020/09/future-propaganda-will-be-computer-generated/616400.
12. Mooney C. The Republican War on Science. New York: Basic Books, 2005.

Literature Cited

13. Koop quoted in: Kolata G. Unwilling to write political prescriptions. New York Times, September 8, 1991, https://www.nytimes.com/1991/09/08/books/unwilling-to-write-political-prescriptions.html.
14. Denworth L. People in Republican counties have higher death rates than those in Democratic counties. Sci Am, July 18, 2022, https://www.scientificamerican.com/article/people-in-republican-counties-have-higher-death-rates-than-those-in-democratic-counties.
15. Molyneux DH, Asamoa-Bah A, Fenwick A, Savioli L, Hotez P. The history of the neglected tropical disease movement. Trans R Soc Trop Med Hyg 115(2) (January 28, 2021): 169–75, doi: 10.1093/trstmh/trab015.
16. Goldenberg S. Michael Mann cleared of science fraud charges made by climate sceptics. Guardian, July 2, 2010, https://www.theguardian.com/environment/2010/jul/02/michael-mann-cleared.
17. Quoted in: McKie R. Climategate 10 years on: What lessons have we learned? Guardian, November 9, 2019, https://www.theguardian.com/theobserver/2019/nov/09/climategate-10-years-on-what-lessons-have-we-learned.
18. Palin S. Sarah Palin on the politicization of the Copenhagen climate conference. Washington Post, December 9, 2009, https://www.washingtonpost.com/wp-dyn/content/article/2009/12/08/AR2009120803402.html.
19. Fischer D. Federal investigators clear climate scientist, again. Sci Am, August 23, 2011, https://www.scientificamerican.com/article/federal-investigators-clear-climate-scientist-michael-mann.
20. Quoted in: O'Grady C. In the line of fire. Science 375(6587) (March 24, 2022): https://www.science.org/content/article/overwhelmed-hate-covid-19-scientists-face-avalanche-abuse-survey-shows.
21. Davenport C, Landler M. Trump administration hardens its attack on climate science. Washington Post, May 27, 2019, https://www.nytimes.com/2019/05/27/us/politics/trump-climate-science.html.
22. Hotez PJ. Preventing the Next Pandemic: Vaccine Diplomacy in a Time of Anti-science. Baltimore, MD: Johns Hopkins University Press, 2021.
23. US House of Representatives. Select Subcommittee on the Coronavirus Crisis. The Atlas dogma: The Trump administration's embrace of a dangerous and discredited herd immunity via mass infection strategy. Staff Report, June 2022, https://coronavirus.house.gov/news/reports/new-select-subcommittee-report-reveals-full-scope-trump-administration-s-embrace.

24. Mallapaty S. Coronaviruses closely related to the pandemic virus discovered in Japan and Cambodia. Nature 588(7836) (December 2020): 15–16, doi: 10.1038/d41586-020-03217-0.
25. Wacharapluesadee S, Tan CW, Maneeorn P, Duengkae P, Zhu F, Joyjinda Y, Kaewpom T, Chia WN, Ampoot W, Lim BL, Worachotsueptrakun K, Chen VC, Sirichan N, Ruchisrisarod C, Rodpan A, Noradechanon K, Phaichana T, Jantarat N, Thongnumchaima B, Tu C, Crameri G, Stokes MM, Hemachudha T, Wang LF. Evidence for SARS-CoV-2 related coronaviruses circulating in bats and pangolins in Southeast Asia. Nat Commun 12(1) (February 9, 2021): 972, doi: 10.1038/s41467-021-21240-1.
26. Worobey M. Dissecting the early COVID-19 cases in Wuhan. Science 374(6572) (December 3, 2021): 1202–4, doi: 10.1126/science.abm4454.
27. Worobey M, Levy JI, Malpica Serrano L, Crits-Christoph A, Pekar JE, Goldstein SA, Rasmussen AL, Kraemer MUG, Newman C, Koopmans MPG, Suchard MA, Wertheim JO, Lemey P, Robertson DL, Garry RF, Holmes EC, Rambaut A, Andersen KG. The Huanan seafood wholesale market in Wuhan was the early epicenter of the COVID-19 pandemic. Science 377(6609) (August 26, 2022): 951–59, doi: 10.1126/science.abp8715.
28. Cohen J. Prophet in purgatory. Science 374 (November 25, 2021): 1040–45, https://www.science.org/content/article/we-ve-done-nothing-wrong-ecohealth-leader-fights-charges-his-research-helped-spark-covid-19.
29. Scudellari M. How the coronavirus infects cells—and why Delta is so dangerous. Nature 595 (July 28, 2021): 640–44, doi: https://doi.org/10.1038/d41586-021-02039-y.
30. Wu Y, Zhao S. Furin cleavage sites naturally occur in coronaviruses. Stem Cell Res 50 (December 9, 2020): 102115, doi: 10.1016/j.scr.2020.102115.
31. Liu X, Wu Q, Zhang Z. Global diversification and distribution of coronaviruses with furin cleavage sites. Front Microbiol 12 (October 7, 2021): 649314, doi: 10.3389/fmicb.2021.649314.
32. Hiltzik M. A Nobel laureate backs off from claiming a "smoking gun" for the COVID-19 lab-leak theory. Los Angeles Times, June 8, 2021, https://www.latimes.com/business/story/2021-06-08/nobel-laureate-baltimore-smoking-gun-for-the-covid-lab-leak-theory.
33. Garry RF. The evidence remains clear: SARS-CoV-2 emerged via the wildlife trade. Proc Natl Acad Sci U S A 119(47) (November 22, 2022): e2214427119, doi: 10.1073/pnas.2214427119.

34. Garry RF. SARS-CoV-2 furin cleavage site was not engineered. Proc Natl Acad Sci U S A 119(40) (October 4, 2022): e2211107119, doi: 10.1073/pnas.2211107119.
35. Rasmussen A, Worobey M. Conspiracy theories about COVID-19 help nobody. Foreign Policy, September 15, 2022, https://foreignpolicy.com/2022/09/15/conspiracy-theories-covid-19-commission.
36. Keusch GT, Amuasi JH, Anderson DE, Daszak P, Eckerle I, Field H, Koopmans M, Lam SK, Das Neves CG, Peiris M, Perlman S, Wacharapluesadee S, Yadana S, Saif L. Pandemic origins and a One Health approach to preparedness and prevention: Solutions based on SARS-CoV-2 and other RNA viruses. Proc Natl Acad Sci U S A 119(42) (October 18, 2022): e2202871119, doi: 10.1073/pnas.2202871119.
37. McDonald J. Republicans spin NIH letter about coronavirus gain-of-function research. FactCheck.org, October 26, 2021, https://www.factcheck.org/2021/10/scicheck-republicans-spin-nih-letter-about-coronavirus-gain-of-function-research.
38. Weixel N. Paul promises investigation of Fauci if Republicans take Senate. Hill, February 24, 2022, https://thehill.com/policy/healthcare/592761-paul-promises-investigation-of-fauci-if-republicans-take-senate.
39. Cohen J. Republican Senate staff tout lab-leak theory of the pandemic's origin. Science, October 27, 2022, https://www.science.org/content/article/republican-senate-staff-tout-lab-leak-theory-pandemics-origin.
40. Orac. NIH funding lies weaponized as disinformation. Respectful Insolence (blog), May 3, 2022, https://respectfulinsolence.com/2022/05/03/nih-funding-lies-as-disinformation.
41. Butler K. Inside the anti-GMO movement's obsession with virology research and lab leaks. Mother Jones, June 14, 2021, https://www.motherjones.com/politics/2021/06/inside-the-anti-gmo-movements-obsession-with-virology-research-and-lab-leaks.
42. Bredderman W. This fave mainstream media source is funded by anti-vaxxers. Daily Beast, November 12, 2021, https://www.thedailybeast.com/us-right-to-know-fave-mainstream-media-source-is-funded-by-anti-vaxxers.
43. Aldhous P. This activist group tapped into partisan COVID politics to make big trouble for Anthony Fauci and the NIH. Buzzfeed News, May 4, 2022, https://www.buzzfeednews.com/article/peteraldhous/white-coat-waste-anthony-fauci-nih-ecohealth-alliance-lab.

Literature Cited

44. Frijters P, Foster G, Baker M. What Covid crimes will victims not forgive? Brownstone Institute, April 14, 2022, https://brownstone.org/articles/what-covid-crimes-will-victims-not-forgive.
45. Stand Up! with Pete Dominick (podcast). Dr Michael Mann and Dr Peter Hotez on combatting the anti science movement. Episode 483, https://standupwithpete.libsyn.com/dr-michael-mann-and-dr-peter-hotez-on-combatting-the-anti-science-movement-episode-482-0.
46. Committee for Concerned Scientists, https://concernedscientists.org/ccs-active-cases.
47. Climate Science Legal Defense Fund, https://www.csldf.org.
48. Union of Concerned Scientists, https://support.ucsusa.org.
49. Hotez P. Who will defend embattled scientists? Boston Globe, January 27, 2022, https://www.bostonglobe.com/2022/01/27/opinion/who-will-defend-embattled-scientists.
50. Southern Poverty Law Center, https://www.splcenter.org.
51. Hotez P. Science tikkun: Science for humanity in an age of aggression. FASEB J 35(12) (November 21, 2021): e22047, doi: 10.1096/fj.202101604.
52. COVID-19 Hate Crimes Act, S. 937, 117th Congress (2021–22), https://www.congress.gov/bill/117th-congress/senate-bill/937/text.

Index

Page numbers in *italic* refer to illustrations.

Ahmed, Imran, 10, 69, 70
alternative medical practices, 29–30, 31
Ambassador Bridge blockade, 36
American Academy of Arts and Sciences, 129
American Medical Association, 31
American Association for the Advancement of Science (AAAS), 128, 129
America's Frontline Doctors, 84
Andropov, Yuri, 113
antibiotics, 97, 108
Anti-Compulsory Vaccination League, 30
Anti-Defamation League, 2, 107, 156
anti-science: anti-Semitism and, 92–93, 107, 156; *vs.* anti-vaccine movement, 9; in authoritarian regimes, 21, 114–15, 116, 117–18, 120–21, 155; disinformation and, 9, 10–11; as global phenomenon, 5–6, 13, 39, 116, 118, 121, 125, 144; government response to, 145–46; historical roots of, 5–6; negative effect of, 5, 19, 141; politics and, 19, 103–4, 134–36, 140; religion and, 146; right-wing groups and, 18–19, 21, 109, 117, 134, 136; rise of, 2, 3, 4, 147–48; social media and, 140
anti-science aggression: challenges of countering, 6, 134, 140–41; definition of, 5, 117; framework for combating, 159–61, *160*; government response to, 159; institutional response to, 158–59; spread of, 3, 154–55
anti-vaccine movement: *vs.* anti-science, 9; anti-Semitic sentiments of, 17–18; attacks on scientists, 10–11, 12–13, 16–18, 87–89; autism narrative

209

Index

anti-vaccine movement (*continued*) of, 7, 9, 14–15; consequences of, 21, 22–23; contrarian intellectuals and, 5, 19–20; COVID-19 pandemic and, 18–20, 22; definition of, 9; disinformation and, 7–8, 10, 68; emergence of, 6, 7; emotional appeal of, 72; ethnic diversity of, 8, 146; far-right extremism and, 15, 18, 117, 144, 146–47; among healthcare professionals, 147; health freedom narrative of, 14, 20, 22, 118; impact on biomedical science, 22–23, 123–27; mass media and, 4, 16, 20, 24–25; Nazi theme in rhetoric of, 91–92, 93; as political movement, 8, 9, 15–16, 20, 34, 73–79, 85–86; social media and, 38, 66, 68, 69, 70, 74; spread of, 7, 8, 35–38, 40, 90, 125

anti-vaccine political action committees (PACs), 8, 15

Applebaum, Anne, 114

Arendt, Hannah: *The Origins of Totalitarianism*, 109

artificial intelligence, 147

Ascent of Man, The (documentary), 133

Atomic Energy Commission, 103

authoritarian regimes: anti-science in, 21, 114–15, 116, 117–18, 120–21, 155; COVID-19 pandemic and, 118–19; vs. totalitarian regimes, 114

autism: causes of, 12, 87; vaccination and, 7–8, 9, 12, 14–15, 32, 87

Autism Science Foundation, 15

Axios (TV show), 20

Azam, Joseph, 41, 64, 83

Baltimore, David, 97

Bannon, Steve: *War Room* (podcast), 18, 100

Baric, Ralph, 151, 153

Ben-Ghiat, Ruth, 116–17; *Strongmen: Mussolini to the Present*, 116

Berenson, Alex, 73; "The Pandemic's Wrongest Man," 85

Beria, Lavrentiy, 112

Berkelman, Ruth, 156

Biden administration: COVID-19 response, 35–36, 53, 61, 78; disinformation advisory board, 86; promotion of vaccination, 73–74

Big Pharma, 33

Bio Farma, 88

Biological E. Ltd., 11, 88

biomedical science: attacks on, 19, 21, 95, 143, 150, 161; in authoritarian regimes, 116; geopolitics of, 1; long-lasting damage to, 123–27; public distrust of, 4; scientific communication in, 6, 140–41

blue states: COVID-19 deaths, 57; excess death rates, 58; vaccination rates, 52–53, 55, 60

Blumenthal, Richard, 97

B'nai B'rith International, 156

Boebert, Lauren, 37, 74

Boehner, John, 74

Bolsonaro, Jair, 115–16

Bonner, Yelena, 113

Bottazzi, Maria Elena, 18, 101

Brady, Kevin, 76

Breitbart (news outlet), 108

Brock, David, 80

Bromley, Allan, 148

Broniatowski, David, 119–20, 136–37

Bronowski, Jacob: *Science and Human Values*, 133

Brooks, Mo, 74–75

Brownback, Sam, 149

Burgess, Michael, 76

Bush, George H. W., 148, 149

Butler, Kiera, 120, 137

Callaghan, Tim, 50, 147
Canada: anti-science in, 36–37, 124–25, 144
Carlson, Tucker, 80, 81, 82; anti-vaccine messages, 18, 80, 81; attacks on scientists, 87, 94, 102, 137; support of authoritarian leaders, 106, 115
Carter, Jimmy, 148
Cato Institute, 137
Caulfield, Timothy, 129
Cawthorn, Madison, 41, 64, 73
Center for Countering Digital Hate (CCDH), 10, 32, 66, 144
Center for Humanities & History of Modern Biology, 134
Centers for Disease Control and Prevention (CDC), 14, 22, 66
Central European University, 115
Cerami, Anthony, 132
China: COVID-19 disinformation, 119
Churchill, Winston, 112
Clarivate Web of Science group, 130
climategate, 149–50
Climate Science Legal Defense Fund, 157, 158
Clinton, Bill, 12
Clinton Global Initiative, 12–13
Cohen, Jon, 94, 151
Cold War: impact on US science, 104; programs in vaccine diplomacy, 121
Comer, James, 152, 153
Committee of Concerned Scientists, 157
Conservative Political Action Conference (CPAC), 72–73
contrarian intellectuals: anti-vaccine movement and, 19–20, 84–86, 134; attacks on science, 5, 6, 103, 104, 112, 116, 137–38; authoritarian regimes and, 109, 116; political support of, 148, 150; public appearances of, 112

Conway, Erik M.: *Merchants of Doubt*, 137
CORBEVAX, 11, 88
Corcoran, Richard, 143
Cornyn, John, 76
coronaviruses: origin of, 96, 97, 150–51
COVID-19 deaths: from Alpha wave, 41, 43, 44; in Brazil, 116; causes of, 100; daily average of, *54*; *vs.* deaths from other diseases, 63; from Delta wave, 41, 43–44, 46–47, 56; in India, 42; among law enforcement in US, 25–26; from Omicron variant, 44, 46; partisan divide in, 20, 53–54, 56, *57*, 58–59; racial disparity in, 48; in Russia, 120; in southern states, 46, 143; in the United Kingdom, 43, 44; US statistics of, *44*; vaccination rate and, 7, 42–43, 45, 47, 48, 53, 143–44; vaccine defiance and, 42, 122, 133; weekly trends in age-standardized incidence of, *45*
COVID-19 Hate Crimes Act, 158
COVID-19 pandemic: in authoritarian regimes, 117–19; hospitalizations, 24, 100; politics and, 94–95, 142–43; preventive measures, 14, 16; public health consequences, 123, 143–44; socioeconomic consequences, 122–23, 144; theories of origin of, 9, 93–94, 96–97, 143, 151, 152–53; unproven treatments, 29, 98, 118, 138, 141
COVID-19 vaccination: disinformation about, 18, 19, 76–77, 112, 119; mandates, 26–27, 28, 74–75, 78, 79; promotion of, 61–62; racial and ethnic equity gap, 48–49; rates of, 46, *55*, 55–56; Russian propaganda on, 118–19; targets for, 43; vaccine development, 11, 14, 88, 91, 101

Index

COVID-19 vaccine hesitancy and refusal: age gap in, 50; emergence of, 20; in Europe, 37–38; geography of, 51–52, 52; mass media and creation of, 4, 64; people of color and, 49–50; political views and, 50–51, 56; reasons for, 5; in Russia, 120; scientists' response to, 62–63; in the United Kingdom, 21–22, 124–25
Covid States Project, 78
Crane, Edward, 137
Cruz, Ted, 94

Dansby, Andrew, 61
Daszak, Peter: attacks against, 97–98, 99, 103, 151–53, 154, 157; public engagement of, 155–56
Davis, Sarah, 34
DC Firefighters Bodily Autonomy Affirmation Group, 27
Delta variant of SARS-2 coronavirus, 25, 26, 41, 43–44, 46–47, 56
Department of Health and Human Services (DHHS), 66–67, 71, 145, 152, 159
Department of Homeland Security, 62, 86, 146, 159
DeSantis, Ron, 77–78, 94, 99–100
Diagnostic and Statistical Manual of Mental Disorders, 12
Dingell, John, 97
diphtheria, 22
DiResta, Renée, 68, 69, 70, 147
"disinformation dozen," 10–11, 12, 15, 32, 66, 70, 85, 159
Doughty, Terry, 79
Douthat, Ross, 59
Dowd, Maureen, 83

Ebola virus, 119
eclectic medicine, 29, 30

EcoHealth Alliance, 96, 97, 98, 152–53, 157
Einstein, Albert, 107, 108
Eisenhower, Dwight, 138, 147
ETH Zurich group, 82, 143
Evangelical Christian groups, 146
excess death rate: COVID/non-COVID, 57, 58; partisan gap in, 60

Facebook: criticism of, 68
Fauci, Anthony: anti-vaccine critics of, 74, 75; attacks against, 93, 94–95, 96, 97–98; death threats to, 25, 87, 95; efforts to discredit, 153; Fox News attacks on, 93–94; lawsuit against, 79; reputation of, 18, 93; request for federal protection, 98; spread of disinformation about, 69, 93–94, 95–96
Faulkner, Harris, 81
Federation of American Societies for Experimental Biology, 158
Feds for Medical Freedom, 27
Feigin Center, 34
fetal tissue research: ban on, 127
Fire Fauci Act, 95
Fiscus, Shelley, 124
Fletcher, Lizzie, 101
Flexner, Abraham, 30
Flexner, Simon, 30
Flynn, Michael, 37
Ford, Gerald, 148
Ford, Matt, 28
Fox News: anti-vaccine messages, 4, 16, 80–82, 112, 143–44; attacks on scientists, 18, 93–94, 102–6, 107, 159; contrarian experts on, 109; health freedom agenda, 79–80; ratings, 83; Russian propaganda machine and, 106; spread of disinformation, 82–83
Freedom Convoy, 36, 37
Freeman, Eric, 49

Gaba, Charles, 20, 47, 48, 52, 53, 56
Gaetz, Matt, 153
gain of function (GOF) research, 96, 97, 98, 152–53
Gamaleya Research Institute, 118
Garry, Robert, 96, 152
Gates, Bill, 27, 39
Gates Foundation, 22, 39, 69, 156
Gavi, the Vaccine Alliance, 22, 125
Gellin, Bruce, 156
George Washington University, 17
Gingrich, Newt, 19, 148
Gladwell, Malcolm, 142
Global Polio Eradication Initiative, 126
Goldwater, Barry, 19, 148
Gorski, David, 84, 139, 153; "Covid-19 and the New Merchants of Doubt," 140
Gosar, Paul, 75
Gray, Lisa, 61
Greene, Marjorie Taylor, 18, 37, 74, 95, 100, 104

H1N1 pandemic flu, 119
Hahn, Stephen, 156
Hamburg, Margaret, 156
Hannity (TV show), 83
Hannity, Sean, 37, 80
Harris, Andy, 75
Health Freedom Defense Fund, 79
Health Freedom for All Act, 26
health freedom movement: campaigns, 14, 20; in Canada, 124; evolution of, 31–32; in Florida, 77–78; global expansion of, 39, 118, *145*; origin of, 28–30, 142; propaganda of, 33, 35, 39, 136; pseudoscience and, 33; in Texas, 34
Health Freedom Protection Act, 32
Heaton, Penny, 156
hepatitis B vaccine, 31
Hesse, Hermann: *Der Steppenwolf,* 122, 141

Hessen, Boris M., 108
HIV/AIDS, 113, 119, 140, 144, 148, 149
Holmes, Dave, 73
homeopathy, 29, 30
hookworm vaccine, 132
Hotez, Eddie, 17
Hotez, Peter: attacks on, 12–13, 66, 88–92, 99–100, 102–3, 105–6; "How the Anti-vaxxers Are Winning," 66; *The Ingraham Angle* interview, 99–100; lectures of, 88–89; Nobel Peace Prize nomination, 101, 102; "The Unique Terror of Being a COVID Scientist after Jan. 6," 95; *Vaccines Did Not Cause Rachel's Autism,* 12, 14, 69
House Un-American Activities Committee (HUAC), 103, 104
"How the Anti-vax Movement Is Taking Over the Right," 86
Hudson, Shanice, 49
human papillomavirus vaccine, 67, 88
Hungarian Academy of Sciences, 114
Hungary vs. Soros: Fight for Civilization (film), 115
hydroxychloroquine, 29, 98, 118, 138, 141

Ibsen, Henrik: *An Enemy of the People* (*En folkefiende*), 101
Imanishi-Kari, Thereza, 97
ImmunityBio, 101
immunization: media portrayal of, 81
Immunization Partnership, 70
IndoVac vaccine, 11, 88
Infectious Disease Society of America, 97
Ingraham, Laura, 18, 81, 99, 102
Ingraham Angle, The (TV show), 83
Ings, Simon: *Stalin and the Scientists,* 111
Inslee, Jay, 26, 60

Institute for Health Metrics and Evaluation, 22
ivermectin, 29, 98, 118, 138, 141

Jackson, Ronny, 76
Jamison, Kay, 129
Jenner, Edward, 138, 139
Joe Rogan Experience (podcast), 85
John Birch Society, 31
Johns Hopkins University School of Medicine, 30
Johnson, Ron, 75
Jones, Phil, 149
Jong-Fast, Molly, 116
Jordan, Jim, 74, 94, 97
JPB Foundation, 15

Kaempffert, Waldemar, 113; "Science in the Totalitarian State," 107
Kennedy, John F., 105
Keusch, Gerald, 152, 155–56
Khazan, Olga, 38
Khrushchev, Nikita, 105, 112
Koch, Charles, 137
Koop, C. Everett, 148
Krugman, Paul, 146

Ladapo, Joseph, 77, 83
Lakshmanan, Rekha, 34
Lamarck, Jean-Baptiste, 110–11
LaMay, Robert, 26
Landau, Lev, 108
Larson, Heidi, 38, 39, 40
Lee, Sheila Jackson, 51
Lenin All Union Academy of Agricultural Sciences, 110, 112
Leonhardt, David, 52, 53
Levit, Solomon, 111
Lincoln, Abraham, 147
long COVID (post-acute COVID-19 syndrome), 123

Lyme disease, 97
Lysenko, Trofim, 110, 111–12

Macron, Emmanuel, 38, 118
malaria, 144, 149
Manhattan Project, 3, 103
Mann, Michael E., 149, 150, 154, 157
March for Science, 128
Masisi, Mokgweetsi, 101
math textbooks: ban of, 143
McCarthy, Joseph, 104
McCaul, Mike, 76
McConnell, Mitch, 76
measles, 7, 10, 13, 22, 31, 32–33, 35, 124–26, 146
measles-mumps-rubella vaccine: autism and, 7, 32, 87; public attitude toward, 7–8, 21–22, 32–33
Media Matters for America, 80, 81, 82, 84, 143
medical education, 30
Mendelian genetics, 108–9, 110–11
Mengele, Josef, 91–92, 109
meningitis, 22
Merchants of Doubt (Oreskes and Conway), 137
Messonier, Nancy, 14
MFG (Menschen Freiheit Grundrechte), 38
Middle American Radicals, 62, 146
Mizelle, Kathryn Kimball, 79
Moderna vaccine, 91
molecular biology, 132
monkeypox, 126–27
Mooney, Chris, 148; *The Republican War on Science*, 19, 147
Moore, John, 124
mRNA vaccines, 14, 18, 45, 48, 68, 91–92
Muller, Hermann, 111
mumps vaccine, 31

Murdoch, James, 83
Murdoch, Lachlan, 82
Murdoch, Rupert, 82, 83, 84
Murthy, Vivek, 66, 67, 68, 79

National Academies of Science, Engineering, and Medicine, 114, 129, 134
National Academy of Sciences, 147
National Aeronautics and Space Administration (NASA), 147
National Center for Immunization and Respiratory Diseases, 14
National Health Federation, 31
National Institute of Allergy and Infectious Diseases, 18
National Institutes of Health, 14
National League of Medical Freedom, 30
National School of Tropical Medicine, 13
National Science Foundation, 114
National Vaccine Advisory Committee, 71
Nature, 90, 98, 130
Nazi Germany: science in, 107, 108–9
neglected tropical diseases, 1, 3, 4, 11, 12, 13, 35, 144, 149
New Abnormal (podcast), 116
New England BioLabs, 156
"New Right," 146
NewsMax, 108
news outlets: anti-vaccine messages, 4, 16; attacks on scientists, 18, 93–94, 99–100, 102, 103; health freedom agenda, 79–82; opinion sections of, 83
NIAID-NIH (National Institute of Allergy and Infectious Diseases of the National Institutes of Health), 69, 93, 96, 102, 143, 152, 153
NIH Fogarty International Center, 156
Nixon, Richard, 148, 149

Nolte, John, 57
Novella, Steven, 140

Oath Keepers, 73
Offit, Paul, 88
Omer, Saad, 62
Omicron variant of SARS-2 coronavirus, 44, 46, 59–60, 79
Operation Warp Speed program, 14
Oppenheimer, J. Robert, 103, 114
Orbán, Viktor, 114–15, 116
Oreskes, Naomi: *Merchants of Doubt,* 137
OSHA (Occupational Safety and Health Administration), 27, 28
Osler, William, 30
Osterholm, Michael, 85, 156

Palin, Sarah, 150
Pan, Richard, 33
parasitic diseases, 88, 132
Paul, Rand, 32, 37, 39, 76, 97, 153
Paul, Ron, 32
Paxton, Ken, 78
pediatric vaccination, 21, 22, 119, 124–25
People's Convoy, 37
pertussis (whooping cough), 22
Pfizer-BioNTech, 91
polio: efforts toward elimination of, 126; global spread of, 10, 22, 125–26; vaccine development, 31
polio vaccinators: assassination of, 40, 126, 146
poverty-related neglected diseases, 11, 88, 101
President's Emergency Plan for AIDS Relief, 149
President's Malaria Initiative, 149
Pringle, Peter, 111, 112
Proud Boys, 12, 28, 73, 134

Public Face of Science Initiative, 129
Putin, Vladimir, 106

Rasmussen, Angela, 152
Reagan, Ronald, 148, 149
Real Anthony Fauci, The (Robert F. Kennedy Jr.), 69
"red COVID," 52, 53, 72
red states: attitude to science in, 4, 65, 148–49; COVID-19 pandemic in, 20, 52–53, 56, 57, 60; mortality rate in, 58, 148; political leadership in, 65; vaccination rates in, 55, 65
Republican Party: anti-science attitudes, 19, 20, 94–95, 96–97, 124, 147–48; religious beliefs and, 62
Requarth, Tim, 122
Research!America, 128
Reyna, Valerie, 72, 129
Rios, Deysy, 49
Roberts, Rich, 156
Rockefeller University Laboratory of Medical Biochemistry, 132
Rolley, Otis, 49
Roy, Chip, 75
rubella vaccine, 31
Rush, Benjamin, 29, 77
Russian propaganda: anti-science and, 144; bots and trolls, 119, 121; on COVID-19, 119; Fox News and, 106; goals of, 119–20; health disinformation and, 118–19, 121; on social media, 119–20; US influencers and, 120; on vaccines, 136
Russia Today (RT) news agency, 119

Sabin, Albert, 31
Sabo, Jason, 34
Sagan, Carl, 139
Sakharov, Andrei, 13, 113
Salk, Jonas, 31, 133

Salk Institute of Biological Sciences, 133
Sarkar, Sahotra, 136
Scalese, Steve, 152
Scheppele, Kim Lane, 115
Scheufele, Dietram, 129
schistosomiasis vaccine, 132
science: human values and, 131–34; public attitude toward, 3–4, 5; as religion, view of, 137–39. *See also* biomedical science
Science, 94, 130, 150
science communication, 6, 127–30
scientific humanists, 133
scientists: anti-vaccine activists' attacks on, 11, 12–13; in authoritarian regimes, 21, 114–15, 116; death threats to, 90, 91, 95, 102–3; far-right attacks on, 16–19, 100, 153–54; mass media attacks on, 18, 93–94, 99–100, 102, 103; media appearances of, 69; patriotism of, 113–14; "pharma shill" label, 11, 18, 33, 89–90; as public enemies, portrayal of, 94–98, 101, 103, 143; public engagement of, 130–32, 155–56; risk-management strategies, 156, 158; social media attacks on, 1–2, 89–92, 98–99; in the Soviet Union, oppression of, 5–6, 13, 104, 106–8, 109–13; surveillance of, 103–4; in totalitarian states, 107–13
Scientists and Engineers for Johnson-Humphrey, 148
Semmelweis, Ignaz, 138
Simons Foundation, 14
Singer, Jeffrey, 137–38, 139
Slaughter, Doug, 49
smallpox vaccine, 138
Smith, Tara, 140
social media: anti-vaccine rhetoric on, 38, 66, 68, 69, 70, 74; attacks on scientists, 1–2, 89–92, 98–99; criticism of,

68; misinformation on, 67–68; Russian propaganda on, 119–20
Soon-Shiong, Patrick, 101
Soros, George, 115
Southern Poverty Law Center, 158
Soviet Union: anti-Semitism in, 106–7; "Doctors' Plot," 106–7; oppression of scientists in, 5–6, 13, 104, 106–8, 109–13; purges in, 21
Sputnik V vaccine, 118
Stalin, Joseph, 5, 106, 107, 108, 110, 111, 112
Stanford Internet Observatory, 70
Stanford University School of Medicine, 102
Stickland, Jonathan, 89, 90
Stokes, Andrew, 57
Straight Outta Merck meme, 88
Swiss Federal Institute of Technology, 82

Tea Party, 14, 74, 142
tetanus, 22
Texans for Vaccine Choice, 15, 34
Texas: anti-vaccine movement in, 35, 89–90; COVID-19 deaths in, 61; health freedom propaganda, 64; World War II fatalities, 61
Texas Children's Hospital Center for Vaccine Development, 11, 101
Texas Medical Center, 30, 34–35, 90, 93, 130, 131
ThisIsOurShot, 71
Thompson, Samuel, 29
Topol, Eric, 156
totalitarian regimes: *vs.* authoritarian regimes, 114; scientific community and, 107–13, 114
Trudeau, Justin, 36
Trump, Donald J., 53, 54, 58–59, 79, 127, 133, 150

tuberculosis, 144
Tucker Carlson Tonight (TV show), 83, 85

Union of Concerned Scientists, 127, 158
unvaccinated Americans: death among, 20, 64–65, 122; demographics of, 48–52
US Food and Drug Administration (FDA), 31, 77, 127
US law enforcement: COVID deaths of, 25–26; vaccine mandates for, 26
US Senate and House Freedom Caucus, 15, 20, 26, 37, 74, 104, 143

Vaccinate Your Family, 70
vaccination: autism and, 7–8, 9, 12, 14–15; decline in rate of, 5; global infrastructure to support, 22; partisan divide on, 124; public defiance of, 5, 8; religious beliefs and, 146
Vaccination Demand Observatory, 40
vaccine advocacy groups, 70–71, 128–29
Vaccine Confidence Project, 38
vaccine-derived poliovirus, 126
vaccine diplomacy, 1
vaccine equity, 49, 71
vaccine-preventable diseases, 22
vaccines: critics of, 24–25; development of, 3, 121; exemptions, 35; hesitancy, 21–22, 38–39; mandates, 26–27, 28, 74–75, 78, 79; passports, 82; for poverty-related neglected diseases, 88, 101; for resource-poor countries, 88. *See also* individual vaccines
Vaccines Did Not Cause Rachel's Autism (Hotez), 69
Vaccines.gov website, 71
Vavilov, Nikolai, 13, *110*, 110–12
Vaxopedia, 70
vernalization: concept of, 110–11

Virality Project, 70
Voices for Vaccines, 70

Walensky, Rochelle, 27
What Covid Crimes Will Victims Not Forgive? (blog), 153–54

White, Jonathan, 49
World Vaccine Congress, 137
Worobey, Michael, 152
Wuhan Institute of Virology, 96, 151

Yamey, Gavin, 140

About the Author

Peter J. Hotez, MD, PhD, is a professor of pediatrics and molecular virology and microbiology at Baylor College of Medicine, where he is also the Texas Children's Hospital chair in tropical pediatrics and codirector of the Texas Children's Center for Vaccine Development. Dr. Hotez is a University Professor of Biology at Baylor University. He is also a fellow in disease and poverty at the Rice University Baker Institute of Public Policy and senior fellow at the Texas A&M Scowcroft Institute of International Affairs and Hagler Institute of Advanced Study. He is an elected member of the National Academy of Medicine and American Academy of Arts and Sciences and the author of more than 650 scientific papers and five single-authored books.

Dr. Hotez has received awards and recognition from the Pan American Health Organization of the World Health Organization, American Medical Association, Association of American Medical Colleges, American Association for the Advancement of Science, American College of Physicians, American Medical Writers Association, B'nai B'rith International, and the Anti-Defamation League, among others. In 2021, he was nominated for the Nobel Peace Prize for his work to combat neglected tropical diseases and to develop low-cost vaccines for global health, including a patent-free COVID-19 vaccine.